# THE WORLD OF FOSSILS

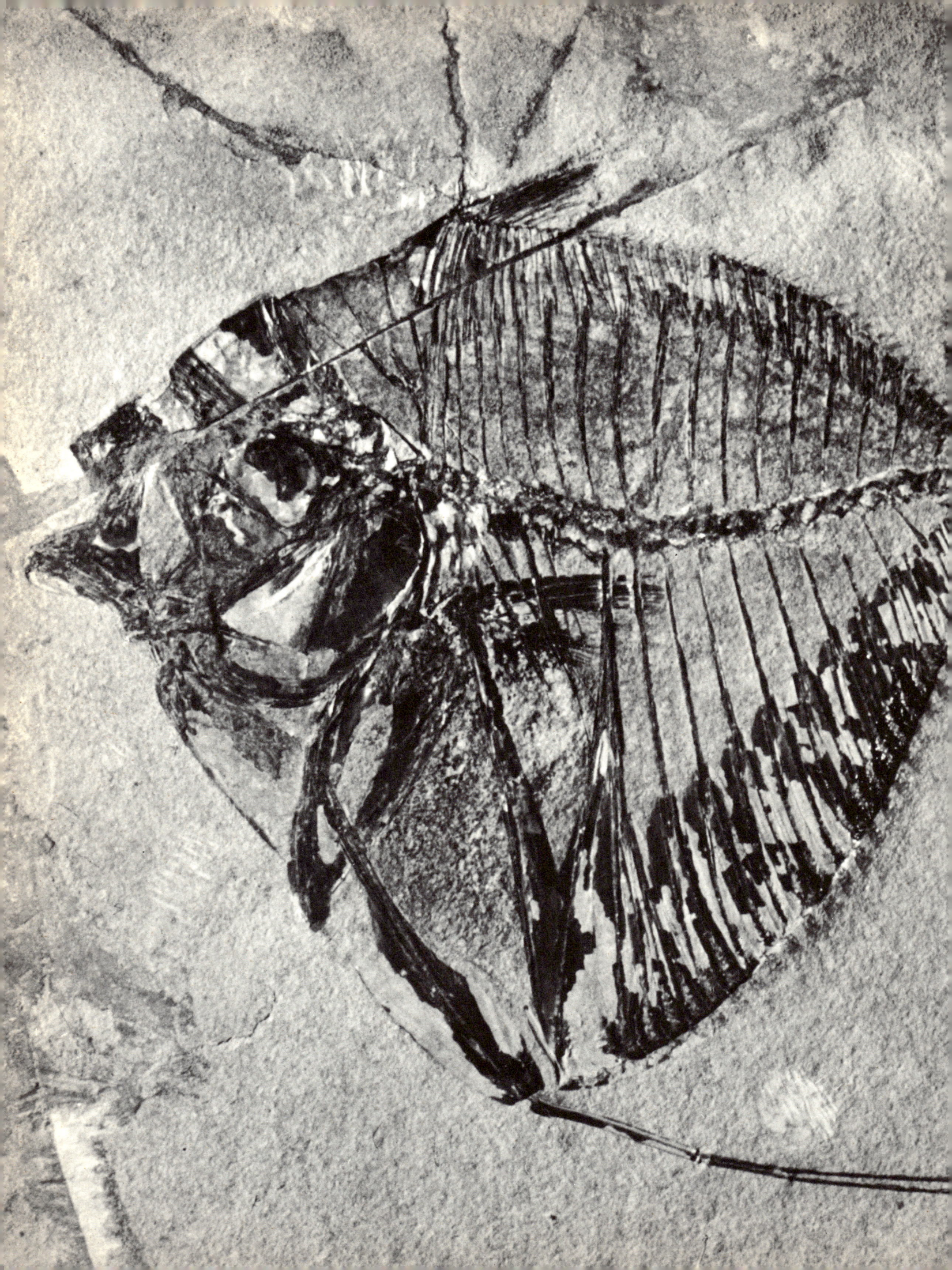

# THE WORLD OF FOSSILS

**GIOVANNI PINNA**
Deputy Director of the Museo Civico di Storia Naturale, Milan

With a foreword by **ERROL WHITE, CBE, DSc, FRS**

Photographs by **CARLO BEVILACQUA**

ORBIS PUBLISHING · LONDON

The author acknowledges the help and co-operation of
Professor Cesare Conci, Director of the Museo Civico di
Storia Naturale in Milan and Dr Giuseppe Pelosio,
Assistant in Palaeontology at the Università di Parma
in the preparation of this book.

All the fossils photographed in this book are from
the collection of the Museo Civico di Storia Naturale
in Milan, except for those on page 9 and page 61
(lower), which are from the Museo di Scienze Naturali
in Verona. The line drawings are by Mario Logli.

*Within recent years there has been a great increase of interest in the life of past ages, and many children, and even some adults, are nowadays familiar with the larger and more spectacular reptiles and other animals that lived long ago, thanks to the imaginative and often excellent restorations that have been used to illustrate popular books, and which have even appeared on postage stamps.*

*However, when it comes to looking for fossils it is not the remains of dinosaurs, mammoths and other big animals that the would-be collector is likely to come across, even in fragments, but those of the simpler, more common forms of life – shell-fish, sea-urchins and the like which may be found, sometimes in abundance, in many holiday places along our coasts and elsewhere.*

*Collecting fossils often starts as a casual pastime to while away idle hours, but many young people develop a craze for collecting, and continue their interest in fossils to become serious collectors, sometimes making a real contribution to our knowledge of animals and plants that vanished long ago. Indeed, the study of fossils is very largely indebted to gifted amateurs who have taken the opportunity and especially the time to collect fossils from favourable localities, a task that can seldom be tackled thoroughly by a professional geologist, who usually has to survey wide areas.*

*As collectors of fossils, young and old, casual or serious, have increased in numbers, so has a considerable literature grown up to meet their needs, usually with the purpose of allowing them to identify their spoils, more or less roughly. In this book we have something rather different from the usual run-of-the-mill books on fossils: its Italian origin gives it a special interest and its illustrations are superb. Yet it is not just another 'coffee-table book' to be flicked through for the sake of the pretty pictures (although these do emphasize the artistic quality of many fossils). The book brings home very clearly one of the chief practical uses of fossils: that of correlating rocks of the same age. In other words, because the succession of faunas and floras has been established, the finding of fossils in widely separated regions can help the geologist to determine the relative ages of rock-formations the world over. In this way, we learn more about the structure of the Earth's crust.*

*The text is written in a general way, avoiding unnecessary detail in the descriptions of the major groups, and laying proper emphasis on what fossils can teach us about the conditions of animal and plant life throughout the ages; and most interesting of all, the evidence they provide of the evolutionary progress of life in general.*

Errol White CBE, DSc, FRS
Formerly Keeper of Palaeontology, British Museum (Natural History)

# Contents

# List of plates

# FOSSILS AND FOSSILIZATION

The word 'fossil' is derived from the Latin meaning 'dug up', or 'extracted from the earth', and was originally used to denote both mineral substances (*fossilia nativa*) and organic substances (*fossilia petrificata*). Only at the beginning of the last century did the term acquire its modern meaning – namely, the remains of an animal or vegetable organism that lived in a past geological era, and has been preserved by means of the process of fossilization.

The science that is concerned with fossils and fossilization is *palaeontology*, 'the study of ancient things'. This is divided into three areas: *palaeozoology*, which is concerned with the animal fossils; *palaeobotany,* the study of fossilized plants; and *palaeoanthropology*, which studies fossilized human remains. Palaeontology is thus an extensive science which touches on zoology, botany, anthropology, and biology, but, unlike those subjects, which are concerned only with the present, it covers the whole succession of life from the time when life first began to develop on our planet.

Fossilization includes all the physical, chemical, and biological processes that bring about the preservation of the organisms inside the rocks of the Earth's crust.

In fact, when an organism dies, the substances that compose its soft parts undergo more or less rapid decay, due to such factors as attack by

*Right: The science that studies fossil plants is palaeobotany, or plant palaeontology. Shown here is a fossil plant from the Eocene layers of Monte Bolca, near Verona, Italy. The height of the specimen is 20 centimetres (8 inches). Far right: The technical name for the study of fossil human remains is palaeoanthropology, or human palaeontology. In the photograph on the right is a cast of the skull of Grimaldi man*

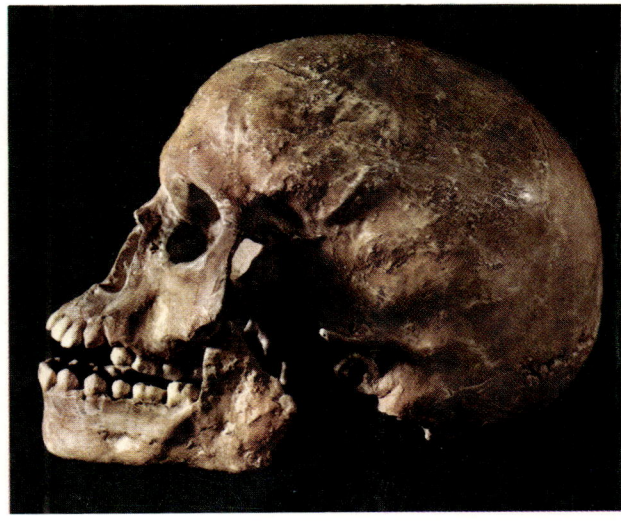

bacteria and erosion by water (particularly the sea), while the hard parts, such as the bone of vertebrates, shells, and so on, may be preserved for a much longer period. If an organism is to be preserved, it must be protected from destructive agents as quickly as possible, and this is what often happens, for example, to a marine organism that falls to the bottom of the sea after its death. It becomes covered by sediment which, once consolidated, protects it from decay. And the sooner that this consolidation occurs, the more likely it is that the organism will be preserved.

The speed of such consolidation naturally varies with the nature of the sediment. In most cases it happens very slowly, and so only 'the hardest and most resistant parts of the organism become fossilized. These parts include the bones, the teeth, the shells, the chitin, the wood, the keratin and the skeletal frameworks of the coelenterates, bryozoa and sponges.

On the Earth there are also certain layers, such as those formed from extremely fine-grained calcareous rocks, which have consolidated so rapidly as to permit the preservation of the most delicate structures of many organisms. This is the case with the now famous Solnhofen stratum in Bavaria, a layer of Jurassic yellow limestone that contains one of the richest and best-preserved fossil faunas in the world. Besides fish and ammonites, there are belemnites complete with tentacles, imprints of medusae, and large insects with their fine membranous wings perfectly preserved. Outstanding among these remarkable fossils is the *Archaeopteryx*, the famous bird-reptile, which was definitively identified as a bird rather than a reptile on the basis of its feathers, which were preserved in the limestone strata. A similar layer, approximately four times as old and perhaps even more interesting, is that of the Burgess clays in British Columbia, Canada. Invertebrates have been found there that date back to the middle of the Cambrian era

*Bones are perfectly preserved in the fossil state: this is the complete skeleton of* Mesosaurus tumidus, *an aquatic reptile that lived in the Permian era about 200 million years ago. São Mateus do Sul, Paraná, Brazil. Length of the original, 25 centimetres (10 inches)*

*Teeth are the part of vertebrates that fossilize best. Below left: A molar of the cave bear,* Ursus spelaeus. *Quaternary; Elba. Below: A classic example of fossilization by mineralization: a section of a silicified trunk of a Triassic conifer. Petrified Forest, Arizona*

(600,000,000 years ago), and which have been of inestimable value to the study of evolution.

These two sites are important because they brought to light fossils that are almost perfectly preserved even in the most delicate structures. Such examples of preservation *in toto*, as the fossilization of the soft parts· is termed, are extremely rare, and are due to certain factors in the process of fossilization. A good example of this is the preservation of insects in amber in certain Oligocene layers of Sicily and of the Baltic. Amber is the fossilized resin of an ancient conifer, the *Pinus succifer*, which in running down

*An example of fossilization 'in toto': the delicate structure of the wings of the 'dragon-fly'* Cymatophlebia longialata *has been perfectly preserved by fossilization. Upper Jurassic; Solnhofen, Germany*

*Near right: Eggs are perhaps some of the most interesting fossil traces. This is an egg of* Protoceratops andrewsi, *a small dinosaur dating from the Mongolian Cretaceous. Height of the original egg is 16.7 cm (about 6½ inches).*
*Right: A Coleoptera fossilized whole, in a fragment of Oligocene amber, Samland, USSR (greatly enlarged)*

Another example of a fossilized insect in the Oligocene amber from Samland (greatly enlarged)

Far left: Neuropteris gigantea, a pteridosperm of the Carboniferous period, whose presence indicates a climate of equatorial type. Dudley, Britain. Length of the specimen is 8 cm (approximately 3 inches). Near left: Leaves preserved in travertine provide a striking example of fossilization by encrustation. Quaternary; Tivoli, Italy

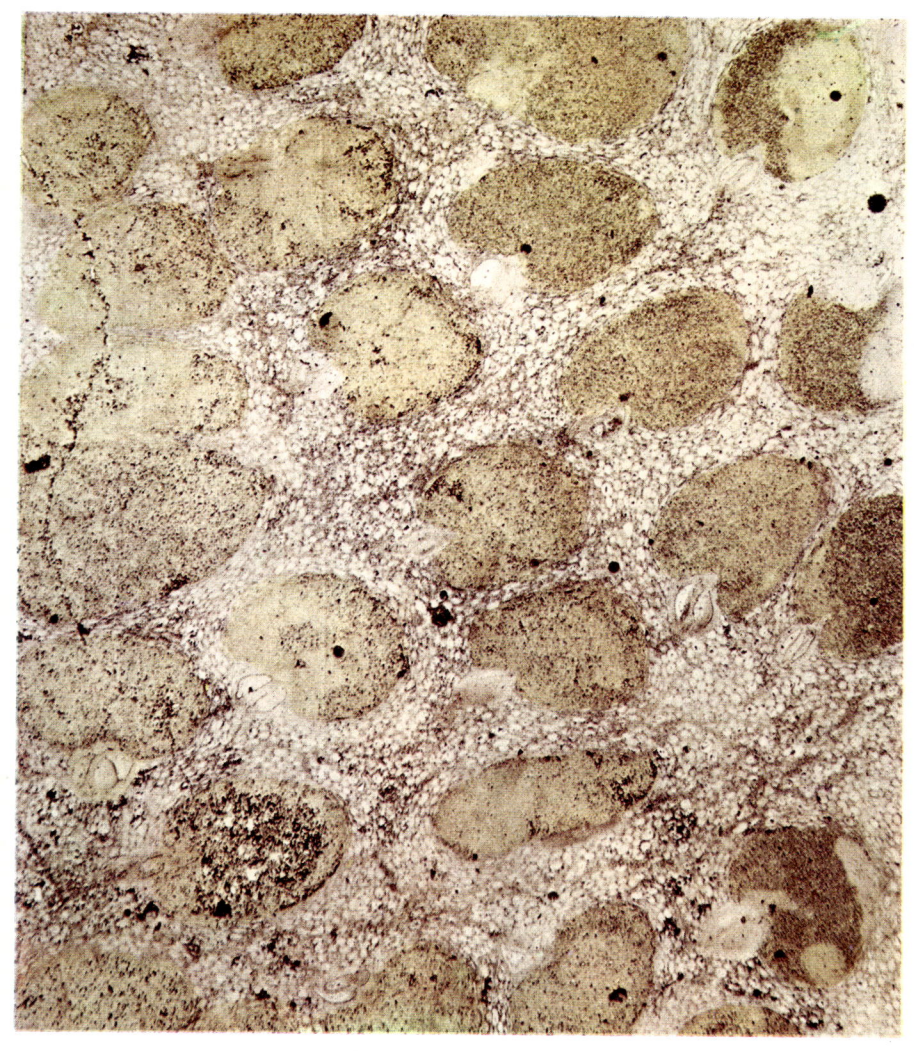

the bark of the tree engulfed various insects, and preserved them perfectly on hardening. Even more spectacular cases of fossilization *in toto* are furnished by the large mammoths of Siberia, found in frozen soil with their skin, flesh and bones intact, and by the mummification of two dinosaurs from the Cretaceous period in Wyoming. These two examples of *Anatosaurus*, 135 million years old, are so well preserved as to show the wrinkled and bruised hide over the abdomen and other parts of the body, as though the corpse had dried out very soon after death, which in turn probably indicates that they lived in a particularly dry climate.

Organisms preserved *in toto* are thus extremely rare. The majority of the fossils that have been found are in fact the result of other types of fossilization: mineralization, carbonization, encrustation, and distillation, to quote some of the most common processes.

The process of *mineralization* is the most common of all, and is the one which more than any other produces typical petrified fossils as they are popularly imagined to be. Mineralization occurs as a result of minerals, carried in the water of the sediments in which the organism is lying, permeating the organism. Sometimes this involves the water merely filling spaces between the skeletal elements, and sometimes a more extensive process, in which all or part of the organism is replaced by mineral substances.

*Above: A thin section of a fragment of a silicified trunk of* Palmoxylon. *The internal structure has been preserved in its smallest detail during the process of fossilization. Quaternary; Libya*

*Right: In this piece of Tertiary rock the fossil lamellibranchs occur in three different forms: some with their original shell, some as internal casts, and some as imprints or external casts. Miocene; Bologna Apennines, Italy*

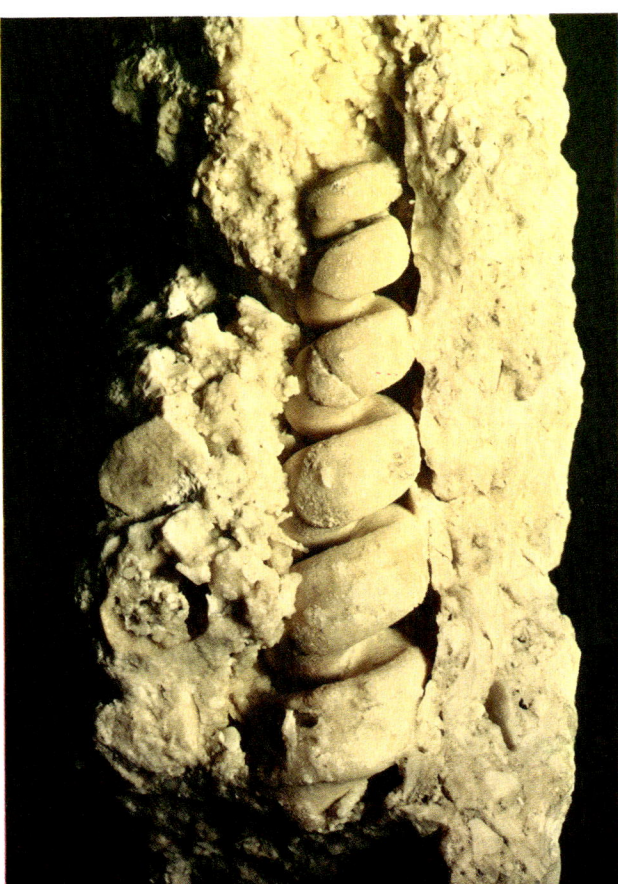

*Far left: A shell in which the organic material has been coarsely substituted by mineral action to produce a pseudo shell. Triassic limestone; Esino, Italy. Height of the original specimen, 5 cm (2 inches). Near left: A perfect internal cast of Turritella cathedralis, a Miocene gastropod from Sardinia (slightly reduced)*

According to the extent of this process, the fossil will either be perfectly preserved, or will be preserved only in a crude general way. A large number of minerals assist in the process of mineralization, but the most common are silica and the calcium salts.

*Carbonization* is a process of fossilization primarily associated with vegetable substances. It occurs as the result of the action of certain anaerobic bacteria (organisms that grow in the absence of oxygen), which break down the organic compounds, liberating gases which are compounds of nitrogen and oxygen, and leaving the organism rich in carbon. It is through this process that the Earth's enormous deposits of coal were formed; the fossilized remains of vegetable accumulations in the soil of a predominantly carboniferous age. In the same way, impressions of leaves, trunks and roots in carbonized form produce important evidence of the luxuriant forests that once existed.

The process of *encrustation*, which is very much less common than either *mineralization* or *carbonization*, generally occurs in terrains more recently in contact with, for example, a chalky spring. In this case the water, loaded with salts, flows over the organism, depositing minerals upon it in the form of a thin encrustation, and the fossil is found in the form of a negative impression. A typical example of encrustation may be seen in the fossil leaves that have been found preserved in travertine, a limestone rock of recent formation.

Finally, *distillation* is the rarest process of all. In distillation the most volatile elements that make up the organism are distilled off, leaving as a residue a thin film of carbon that takes the form of the object.

Thus, many different types of fossil may be found. For example in the case of a fossil spiral shell, the substitution, molecule by molecule, of minerals for the substance of the organism results in the original shell being perfectly preserved, down to the smallest details of its structure. In a cruder process of mineralization, the shell is replaced by the mineral only in its general outline, and what is then seen is a simulated shell. In some cases too, the shell has completely disappeared, leaving its inner parts exposed, and this is known as an internal cast but is naturally often difficult to classify. Finally, an external impression is obtained when an organism rests upon the sediment, leaving an imprint.

Thus, the principal means of formation of fossils are mineralization, carbonization, encrustation and distillation; it is on these physical events that much of our knowledge of the past relies. And not only are fossilized organisms found; as we shall see, records of animal traces, droppings and so on are often valuable.

# THE SCIENCE OF PALAEONTOLOGY

As we have said, palaeontology is the science that is concerned with fossils. Its principal aim is to achieve – through the study of fossils – the most accurate possible reconstruction of the geographical and biological history of our planet, of the organisms that have lived on it, of their evolution, of their way of life, and of their relationships with each other.

*Palaeoclimatology*, the subdivision of palaeon-

ancient epochs. For example, as a result of the discovery of vast deposits of coal in many countries of the Northern Hemisphere, formed from forests corresponding to the present equatorial rain forests, scientists believe that during the Carboniferous era (250 million years ago) these regions had a hot humid climate of the type found nowadays only at the equator. In a similar way, the discovery of fossil corals in

*Two Quaternary molluscs that are used as geological thermometers in determining fluctuations of climate during the Quaternary era:* Mya truncata *(left) indicates a cold climate, while* Conus testudinarius *(right) indicates a warm climate. From Quaternary rocks in Sicily*

tology that is concerned with the investigation of the climates of prehistoric eras, making use of fossils as geological 'thermometers', is an essential side of this work. Starting as it does from the principle that the fossil organisms underwent modifications and evolved in a similar way to analogous organisms now living, it can establish, by means of the comparative study of fossils, the climatic conditions of the place where the fossil was found, during the period when that fossil was alive.

Such climatic reconstructions are of course much more easily carried out for periods that are not too far removed from our own times, since comparisons between living and fossilized species are easier to make. Nevertheless, they have also been attempted for some of the most

*Above: Two Quaternary hot occupants:* Strombus bubonius *(left and* Patella ferruginea *(right), from Taranto and Sicily respectively*

*Left: The presence of some types of algae calcified in marine sedimentation indicates the existence of a warm climate at the time of deposition. Shown here is* Sphaerocodium kokeni, *Middle Triassic; Heskem (Germany)*

the Triassic dolomitic rocks of the Alps suggest that the Triassic period enjoyed a warm and humid climate.

Molluscs are also climatic indicators. In this connection the study of the lamellibranchs and gastropods of the Quaternary period is extremely interesting. In the Quaternary period, which saw wide fluctuations in climate from glacial to hot interglacial periods, numerous species of mollusc today found in northern seas migrated into the Mediterranean in the cold periods, while yet others went during the hot interglacial periods to the coasts of Senegal, where they still live today.

In the course of his work, the palaeontologist turns to *palaeoecology*, the study of the environment of past eras, which is concerned in particular with the comparative dating of rocks and the study of the evolution of past organisms. Again, comparisons between fossils and living

*Below:* Platax *sp., the finest fish found in the Eocene layers of Monte Bolca, Verona, Italy. It indicates a tropical climate*

| \multicolumn{4}{c}{**Order of the Geological Eras**} | | | |
|---|---|---|---|
| Era | Period | Epoch | Age (Millions of Years) |
| Cenozoic (Upper Tertiary and Quaternary) | Neogene | Holocene | 1 to Recent Aluvium |
| | | Pleistocene | 4—1 |
| | | Pliocene | 11—4 |
| | | Miocene | 25—11 |
| | Palaeogene | Oligocene | 40—25 |
| | | Eocene | 70 —40 |
| | | Palaeocene | |
| Mesozoic | Cretaceous | Upper Cretaceous | 135 —70 |
| | | Lower Cretaceous | |
| | Jurassic | Upper Jurassic | 180 —135 |
| | | Middle Jurassic | |
| | | Lower Jurassic | |
| | Triassic | Upper Triassic | 225 —180 |
| | | Middle Triassic | |
| | | Lower Triassic | |
| Upper Palaeozoic | Permian | Upper Permian | 270 —225 |
| | | Lower Permian | |
| | Carboniferous | Upper Carboniferous (Pennsylvanian) | 350 —270 |
| | | Lower Carboniferous (Mississippian) | |
| | Devonian | Upper Devonian | 400 —350 |
| | | Middle Devonian | |
| | | Lower Devonian | |
| Lower Palaeozoic | Silurian | Upper Silurian | 440 —400 |
| | | Lower Silurian | |
| | Ordovician | Upper Ordovician | 500 —440 |
| | | Middle Ordovician | |
| | | Lower Ordovician | |
| | Cambrian | Upper Cambrian | 600 —500 |
| | | Middle Cambrian | |
| | | Lower Cambrian | |

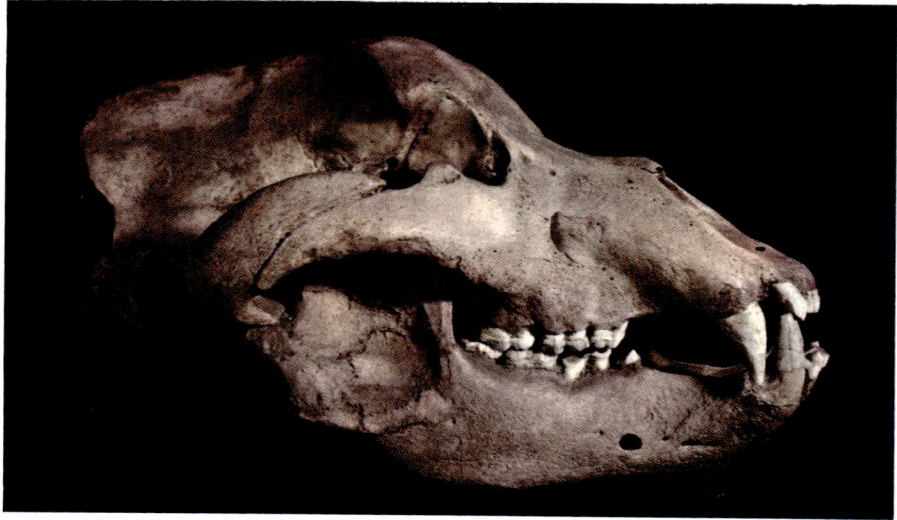

species form the basis of his work. It is, in fact, possible to determine from fossils the conditions in which the fossils and the rock that contains them were deposited, and thus to learn something about the environment. For example, the finding of a frog might indicate surroundings of a marshy nature, or a lake; the teeth of a vertebrate land animal – the proximity of land; corals – sea and reefs. Again, the most detailed of such indications are furnished by molluscs, which contain clues from the past: whether salty or fresh water, and deep or shallow sea.

A final important problem resolved by palaeontology is that of the relative dating of the terrestrial rocks. In geology, where time is measured in millions of years, there are two possibilities for dating rocks: one is the absolute method that establishes exactly the number of years that have passed since their formation; and the other is the relative method, which determines only whether a rock is older than, contemporary with, or more recent than, another. This last system is based on the concept that each epoch of Earth's history has had characteristic animal and vegetable types, which are not comparable with those in any other earlier or later epoch, and on the basis of a study of fossils it is thus

*Mene rhombea, another fish from Monte Bolca that indicates a tropical climate, Length 19 cm (about 8 inches)*

*The skull of a typical land vertebrate from a cold climate, Ursus spelaeus. Quaternary; Grotta di Pocala, Trieste, Italy*

possible both to arrive at the position that they occupied in geological time, and to ascribe an age to the rock that contains them.

It is this principle that provides the basis for the concept of the 'guide fossil', that is, an animal or plant species belonging to the groups with a rapid evolution which have had a short geological life, and thus a limited vertical distribution among the rocks, but a distribution over a very wide area. It is by means of such organisms that correlations are established between rock strata of the same age but located at considerable distances from one another.

On the basis of these 'stratigraphical' studies it has been possible to subdivide the history of the Earth into different eras, each lasting many millions of years.

In the table on page 9 (which by convention

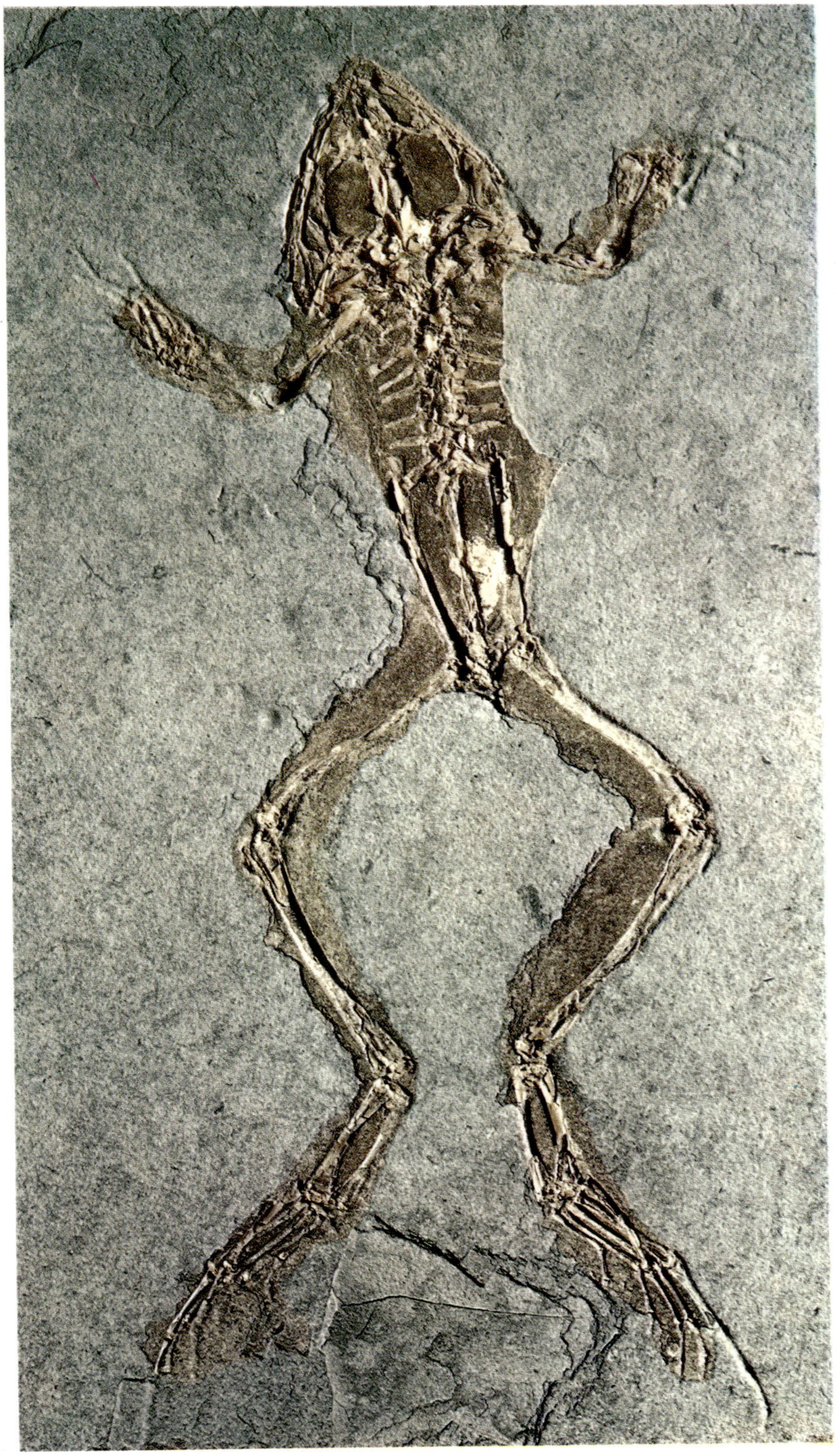

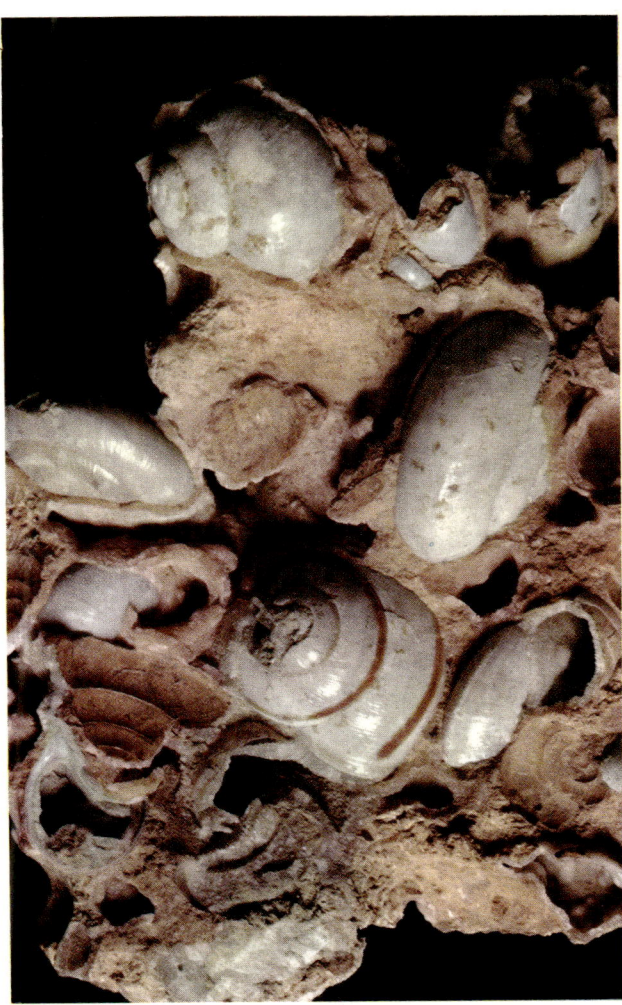

*Left: A sure indication of marshy conditions:* Rana pueyoi. *Upper Miocene; Spain. Length 20 cm (8 inches).*
*Above: Spirals of the terrestrial gastropod* Helix. *Quaternary; Zandobbio, Bergamo, Italy ( × 1.5)*

should be read from the bottom upwards), the major eras and their subdivisions are given beside their approximate ages.

With the ability to date the terrestrial rocks, and to establish the climatic and ecological conditions in the various epochs, the palaeontologist is in a position to carry out palaeogeographic reconstructions; that is, to establish often with considerable precision, the relative disposition of the earth masses and the seas.

A classic example is the reconstruction carried out in Northern Italy on the Pliocene period (10 million years ago). Examination of Pliocene layers, which are made up of sands and clays containing coastal molluscs, furnished evidence of the existence of a broad gulf occupying more or less the position of the present Po Valley, stretching from the Alps in the north and to the west, and to the Apennines in the south.

As might be expected, the work of the palaeontologist has made an enormous contribution to the study of evolution. Palaeontological discoveries are in fact among the most important proofs of evolution itself and have made it possible to establish and confirm the relationships between the great animal groups. In certain cases, important finds have enabled scientists to follow the evolution of entire groups step by step, as in the cases of horses and elephants, the remains of which are often discovered in Quaternary deposits.

*Seymouria baylorensis, a vertebrate of the Permian era which is considered to be the evolutionary link between amphibians and reptiles. Texas. Length of the skeleton 60 cm (24 inches). A cast of the original is preserved at the US National Museum in Washington*

# INVERTEBRATE FOSSILS

Before embarking on any detailed discussion of the fossil world, however, a few words should be said about *systematics*, the science of classification of living organisms, first elaborated by the Swedish scientist Linnaeus, in the eighteenth century. The fundamental units of classification are the species, which are grouped together in genera, which in their turn are classed into families, and so on. This system is applied to both the plant and the animal kingdoms, the full list of subdivisions running as follows: species, genus, family, order, class, phylum, kingdom. Each animal or vegetable fossil has a double Latin name, of which the first part represents the genus and the second part the species to which it belongs. (The letters *sp.* placed after a generic name indicate that the species is undetermined.)

The name of any given species may have many origins: it may be derived from the name of the man who first described or discovered the species, or from the locations in which the species is found, or may describe some feature of the organism, such as its colour, shape, or habits. Modern classification is still, as in Linnaeus' day, based on the comparative structure of plants and animals, but thanks to the discoveries of modern science it is now much more exact than formerly. Even so, classifications change, sometimes causing considerable confusion for the amateur. In this book we have attempted to give the most up-to-date information, although in some cases this has been simplified.

## The Protozoa

At the start of a book of fossil invertebrates which is intended for both classifiers and collectors, it is appropriate to devote a few lines to a group of organisms that are not easily seen in rock layers because of their small size, but which are to be found in large numbers in the sedimentary rocks from all eras and in all countries. These are the Protozoa, primitive unicellular organisms whose study is very different from that applied to other categories of fossils. Protozoa are the concern of micropalaeontology, the science that covers all microscopic fossils. Micropalaeontology is not limited, however, to the study of protozoa alone, since sedimentary rocks frequently also yield small algae, spores, minute crustaceans and microscopic remains of other organisms such as fragments of sponges, sclerites of holothuriae, fragments of bryozoa, and mandibles of annelids. Thus micropalaeontology embraces a vast field, rendered still more complex by the technical details with which it is involved, and which vary according to the type of research, and the nature of the rock or of the organism that it contains.

Fossil protozoa are found in all types of sedimentary rocks. The methods used for collecting them vary according to whether the rocks involved are loose, such as clays or sands, slightly coherent, such as marly limestone, or very solid, such as limestones, dolomites and sandstone grits. In the first case, simple washing in a sieve under a jet of water is sufficient to extract the microfossils from a specimen. This separates the rock to leave a residue from which, when dried, the fossils are extracted and examined under a microscope. The slightly coherent rocks, on the other hand, must first be broken down and then boiled in oxygenated water. When this process is completed, the washing is carried out as in the previous case.

Both techniques free the microfossils completely from the substance that enclosed them, and after being placed in special holders the samples may be studied under the microscope. In the case of the more solid rock which cannot be easily broken down, the method of research and study alters appreciably. Extremely thin sections of rock are prepared, much in the same way as a histological section, and examined under the microscope. Since the sections are so thin, however, the palaeontologist has to take several of the same species in order to make a reconstruction of the shape and structure of the organism.

The protozoa are principally aquatic animals, widespread in the sea and in fresh water, where they live freely and independently in large quantities, sometimes in isolation, sometimes grouped in colonies. Their body is made up of a single cell capable of performing all vital functions, from reproduction and movement to nutrition and defence. They represent the lowest rung of the evolutionary ladder of the animal kingdom, and are considered to be its oldest representatives.

The protozoon, with its simple structure,

*Polished rock surface showing numerous sectional specimens of Neoschwagerina craticulifera, fusulinids of the Upper Permian. Japan. Diameter of the largest specimen, 0.7 cm*

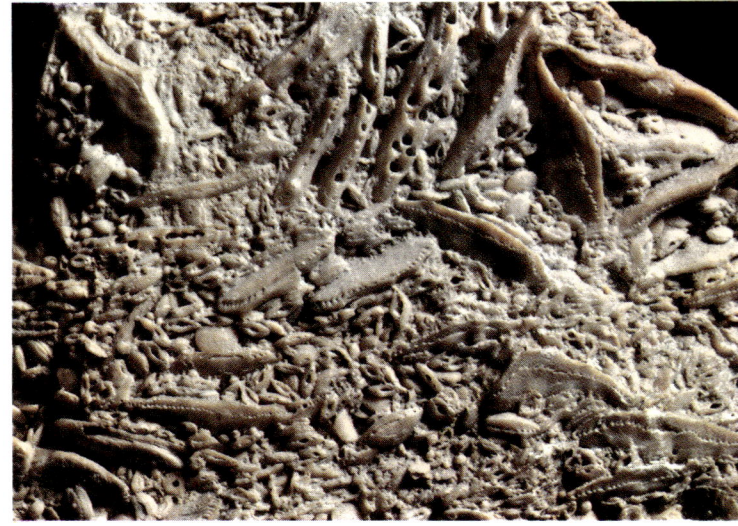

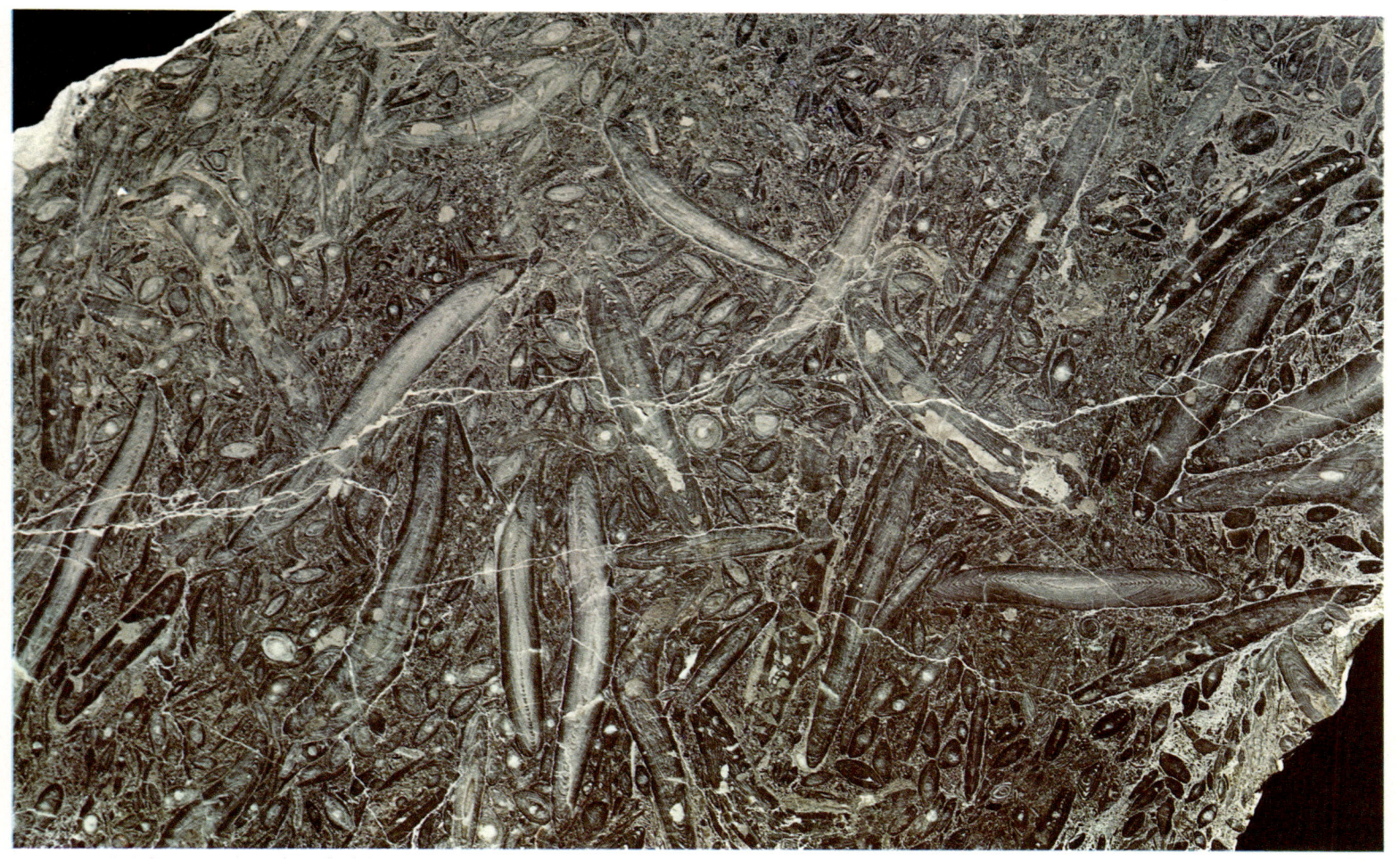

consisting as it does of cytoplasm and one or more nuclei, is easily destroyed, and as a result good fossilization of such organisms is rare. In fact, the only fossils that have been found are those that have been at least partially preserved because they possessed a solid outer shell. Such specimens include the radiolaria, which possess a silica shell, and the foraminifera, with a calcareous or chitinous shell. These two orders, together with other less important groups, are placed under the classification of rhizopods.

### Radiolaria

The radiolaria are protozoa that live today, as in the past, deep in the oceanic seas. They consist of a single cell enclosed in a silica shell, spherical, oval or lenticular in shape, closely perforated, in which spines and long prickles are inserted. The importance of these organisms, which are known to have existed as early as the Cambrian period (500 million years ago), lies in the fact that over the course of time they were deposited as stratified siliceous rock (radiolarites), formed from the consolidation of the 'radiolarian mud'. This is still to be found today, at the bottom of the Indian and Pacific Oceans, as the result of accumulation of radiolarian shells. Radiolaria of the Jurassic period have been found on brightly

coloured outcrops in the Alps in Lombardy, and are also to be found in the Eozoic rocks of France, the Devonian of Germany, and elsewhere. They are particularly abundant in the late Cretaceous and early Tertiary rocks of California.

### Foraminifera

The foraminifera, certainly the most common of all the fossil protozoa, have the simplest structures of all. They consist of a single cell possessing one or more nuclei, from which thin projections known as pseudopodia extend, and they are enclosed in a shell that may take many forms. It may be chitinous, calcareous or agglutinated. In the last case it is made up of minute particles of different substances cemented together; it can be unperforated, with one single large aperture, or alternatively perforated, with the whole of its surface covered in fine holes through which the pseudopodia of the cell protrude. The shape is also very variable: in one type (monothalamous foraminifera), the shell is composed of one single chamber of a globular, lenticular, elongated, stellate or spiral form, and in others (polythalamous foraminifera) there are a number of chambers arranged in single, double, or triple series, either spirally or concentrically arranged.

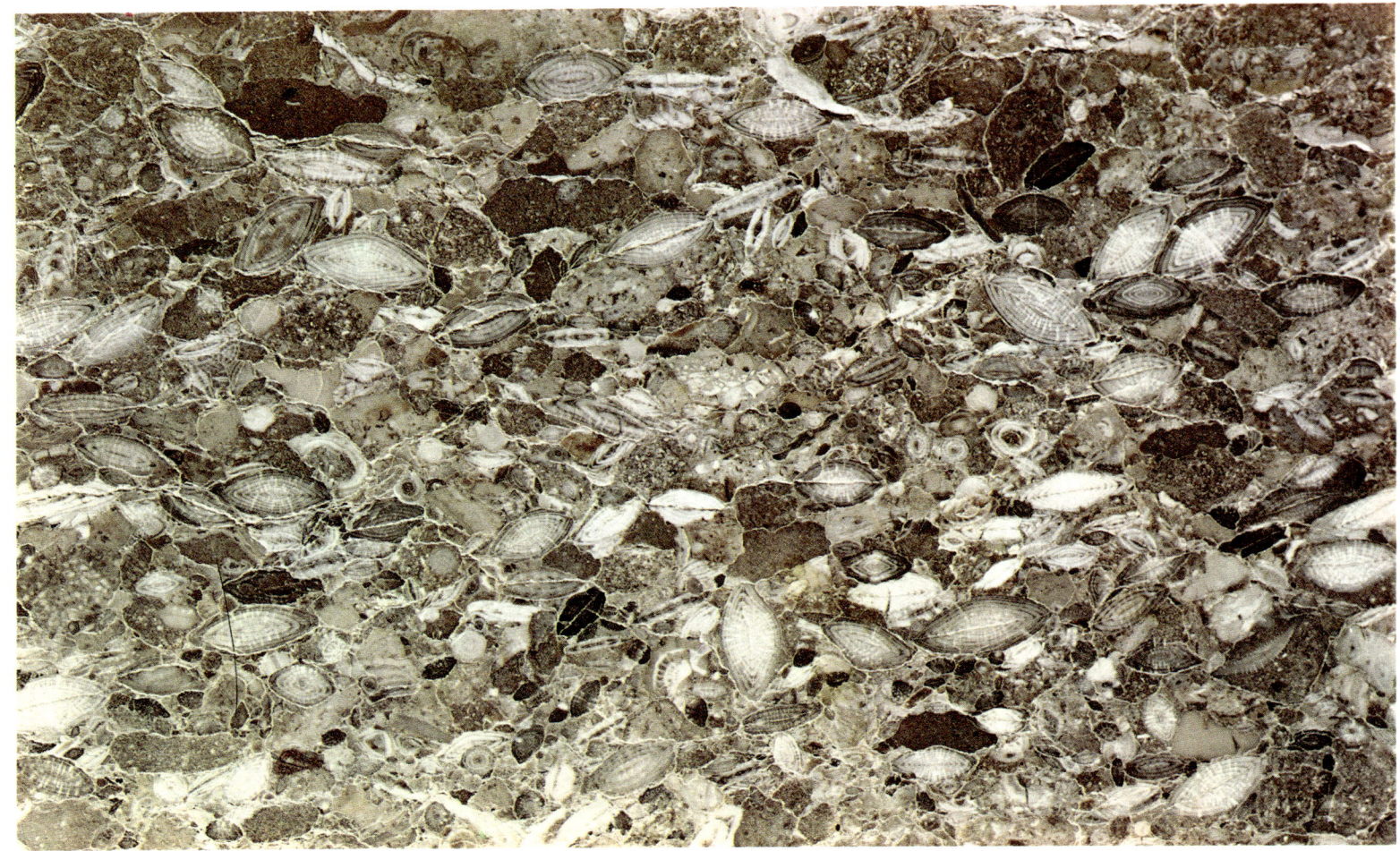

*Limestone containing alveolinae. Middle Eocene; Carso, Trieste, Italy*

The foraminifera are adapted to many different types of marine habitat, ranging from the oceanic to the coastal; one single family, the Allogromidae, unknown in the fossil state since it lacks a shell or has only a chitinous shell that is easily destroyed, lives in fresh waters.

The importance of these fossils to palaeontologists is twofold. First of all, large numbers of fossils are extremely sensitive to the conditions of their environment – notably to salinity, temperature, the depth at which they lie and the type of the bottom, and each species is linked to a particular set of ambient conditions. It can therefore be seen that useful indications regarding the conditions of the formation of the rock in which they were found can be gained from comparisons with the living species. The second factor that makes the study of this animal group extremely important is the presence of large numbers of different species dating from the Cambrian era to the present day, each one of them characteristic of a different fraction of time. These can serve as useful guide fossils in the dating of the strata in which they are found.

Leaving aside the countless protozoan families, which principally consist of organisms varying in size from 0.01 to 1 millimetre, we will consider here some of the much larger fossils which have diameters of as much as 10 to 15 centimetres (4 to 6 inches). These macrofossils are known collectively as macroforaminifera.

Of these, the fusulines are the oldest. They appeared during the Carboniferous period, were developed through the Permian and were extinct by the end of this period, after a life of about 70 million years. Of all the palaeozoic foraminifera, these are the largest and the most complex in structure. Their shell is imperforate, in the shape of an elongated shaft that twists on its axis, inflecting at regular intervals to form a large number of elongated internal chambers, each inflection corresponding on the external surface to a clearly visible longitudinal furrow. The fusulines are found principally in calcareous rocks, where they are frequently associated with algae, and are thought to have lived in shallow, warm seas with limpid water.

The nummulites, the last group that we shall consider, are well known to the collector, and are in fact the only foraminifera to be mentioned by historians. It is recounted that Ancient Egyptian priests, finding large numbers of these small round shells on the sides of the pyramids of Gizeh, were convinced that what they were seeing were petrified lentils, residues of the food of the slaves employed in the construction work.

The pyramids are in fact constructed with blocks of Eocene limestone rich in fossils, and represent one of the most imposing deposits of nummulites. The name *nummulites*, meaning 'little coins', came later, as people associated their round shape with pieces of money.

Many species of the nummulite group were alive in the Eocene period, about 60 million years ago. The shell, which in larger specimens may be as much as 12 centimetres (5 inches) in diameter, is made up of an external wall of two calcareous layers, which turns in a spiral about its axis, progressively increasing in size. Inside, the shell is divided into a number of chambers, formed by regular inflections of the lower layer of the wall, and communicating with each other through a small hole. They are benthonic organisms, and probably lived at depths between 150 and 450 feet on calcareous or sandy sea beds.

A classic site at which these fossils have been found is in Italy, in the pre-alpine region between Verona and Vicenza. They are also commonly found in rocks of Mesozoic and Palaeozoic age wherever there are marine deposits. They are used as important horizon markers in Cretaceous rocks from the western United States.

# The Sponges

Protozoa possessing a *flagellum*, a long filament that serves to propel the organism, are known as flagellate protozoa. Some of these occur in colonies made up of numerous individuals, which are sometimes identical and independent of each other, and sometimes different and with different functions, according to the position that they

| Phylum | Class | Order | Age |
|---|---|---|---|
| Porifera or Spongea | Demospongea | Keratosida | Carboniferous-Recent |
| | | Haplosclerida | Cambrian-Recent |
| | | Poecilosclerida | Cambrian-Recent |
| | | Hadromerida | Cambrian-Recent |
| | | Epipolasida | Cambrian-Recent |
| | | Choristida | Carboniferous-Recent |
| | | Carnosida | Carboniferous-Recent |
| | | Lithistida | Cambrian-Recent |
| | Hyalospongea | Lyssakida | Lower Cambrian-Recent |
| | | Dictyida | Middle Ordovician-Recent |
| | | Lychniskida | Trias-Recent |
| | | Heteractinida | Lower Cambrian-Carboniferous |
| | Calcispongea | Solenida | Cambrian-Recent |
| | | Lebetida | Lower Jurassic-Recent |
| | | Pharetronida | Permian-Recent |
| | | Thalamida | Upper Carboniferous-Cretaceous |

occupy within the colony. This type of organization suggests a possible evolutionary thread towards the formation of multicellular animals (metazoa) made up of tissues having differentiated cells. Precisely when such a transition might have occurred, however, and how, is one of the still unsolved problems of the history of the animal kingdom. Nevertheless animals exist today that are very similar to those found in the oldest fossil-bearing strata (Precambrian period), which could in many ways be considered to be intermediate between the protozoa and the metazoa. These are the porifera, or sponges as they are better known, which have a higher cellular organization than colonies of protozoa and are thought to have come from the primitive flagellates, because they possess microscopic simple cells equipped with a long flagellum.

Thus, the sponges approach the subkingdom of the metazoa. For although the sponges do not possess cells organized into tissues, such as can be seen in the more evolved invertebrates, they nevertheless present three well-differentiated types of cell. These are: concealed epithelial cells that cover the external surface of the colony, flagellate or choanocyte cells that line the cavities and the internal ducts of the body, and wandering amoeboid cells, in the substance of the organism, which are able to live outside the sponge, even if only for a few days. Besides the partial individuality of the cells, the features that associate these animals with the protozoa are the ability of the organism to regenerate itself from a fragment or, even more striking, the ability of different individuals to merge themselves into a single unit.

Whether protozoa or metazoa, the sponges are aquatic, animals with a body that has no precise shape. Modern species exhibit three principal types of internal organization. The simplest type of sponge (the asconoid) is simply a soft bag that lives attached to the sea bed by its lower part. Its surface is covered with small pores through which water passes, into an internal cavity known as the simulated gastric cavity where the exchange of gases and nutrients

*Left:* Protospongia rhenana, *a siliceous sponge of the Lower Devonian. Bundenbach, Germany. Length of the original, 20 cm (8 inches).*
*Below:* Raphidonema farringdonense, *a calcareous sponge perfectly preserved in British Lower Cretaceous strata*

Synolynthia *sp., an organism resembling a calcareous sponge. Cretaceous; Foots Cray, Britain (slightly enlarged)*

*Right: Schematic drawings of three forms of internal structure of a sponge: A. Ascon; B. Sycon; C. Leucon; Ac. Afferent canal; Ec. Efferent canal; Fc. Flagellate chambers; Pc. Pseudo gastric cavity; Os. Osculum. Far right: Sponge spicules: A. Monoaxon; B. Triaxon or hexactine; C. Tetraxon; D. Triaenes*

takes place. This is lined with choanocytes, equipped with flagella which force the water out of the body through an upper aperture or outlet mouth. In the second type (the syconoids) the organization is almost identical, but the simulated gastric cavity is regularly inflected to form niches lined with choanocytes, which provide a much larger surface area over which exchanges may take place. The third, most evolved type (the leuconoids) is that in which there is a genuine flow system, with inflow ducts to convey water into the organism and to a chamber lined with choanocytes, and outlet ducts to force it into the simulated gastric cavity, which is considerably smaller in size. From there it is expelled through the outlet mouth.

This soft structure, which is easily destructible, and is only rarely preserved in the fossil state, is supported by an internal skeleton made up of a system of calcareous or siliceous spicules (megasclerites) joined together to form a reticulated bonding, which can sometimes be seen by the naked eye. In addition to these are a number of other types of spicules distributed throughout the interior of the body, which are very much smaller and visible only under the microscope. These are known as microsclerites, and are often used, along with the choanocytes, to assist zoological classification. On the death of the animal, however, these structures disperse, so that the classification of fossil sponges is based upon the type and upon the shape of the megasclerites and upon the network that they form.

The study of the spicules, which are occasionally found dispersed in the sedimentary rocks in such large numbers as to form the siliceous rocks known as 'spongolites', is again included in the field of micropalaeontology, and is chiefly carried out with the aid of a microscope. Micropalaeontological studies of sponges have revealed the existence of different types of spicule: the siliceous type (crystallized opal or chalcedony), which has a small central duct that can be clearly seen in the fossil, and the calcareous type, in which that duct is rarely preserved. The most simple type of spicule is the monaxial, a long needle formed on a single axis. More complex types include the triaxial or hexatine types of spicule, which have three axes perpendicular to each other, the tetraxial spicules with four axes, and the triatine spicules with three axes which are found in the same rock layers and are characteristic of the calcareous sponges.

The importance of the sponges to the palaeontologist lies principally in their use as guides to the environment in which the animals lived. For their forms vary according to the nature of the currents, the clarity of the waters and their depth, so that the fossil sponge species give us important clues about the conditions existing at the time when they were alive.

Fossil sponges have been found dating back

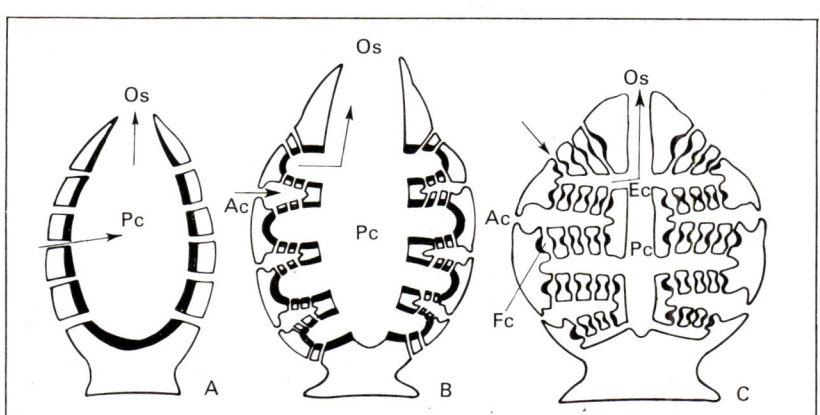

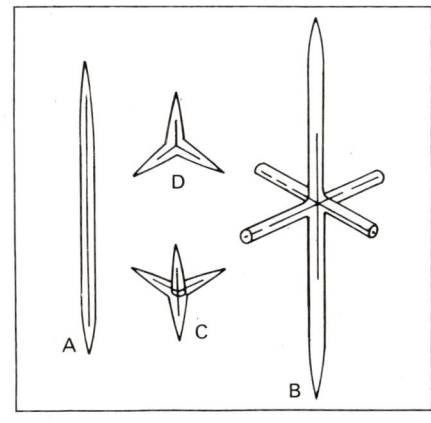

to the earliest geological eras. In the oldest rocks of the Archaeozoic era, which generally have very few fossilized remains, evidence of the existence of sponges has been found, and this many millions of years before the great explosion of life which took place at the beginning of the Palaeozoic era. The problematic fossil *Atikokania*, a cylindrical or conical sponge discovered in 1912 in Precambrian terrain in Canada, is without any doubt the oldest known representative of the group, dating back about 1,300 million years. Sponge spicules originating from the Precambrian rocks of the Grand Canyon, a mine of many of the oldest fossils, have also been found; and in rocks in Zaire (formerly the Congo), organisms have been found which are similar to present-day calcareous sponges, while calcareous spicules found in Brittany are attributed to a very old sponge called *Eospicula*.

The history of the evolution of the sponges, from the beginning of the Palaeozoic era up until the present day, has changed very little: the sponges have in fact been preserved virtually unchanged through the course of the geological eras for over 500 million years. And as early as that, the phylum Porifera included representatives of the three classes into which they are still divided: Demospongea, to which the 'bath

*Above:* Receptaculites *sp., an unusual Palaeozoic organism thought to be related to the sponges through its structure, and the shape of the numerous canals that cover its surface. Devonian; Germany*

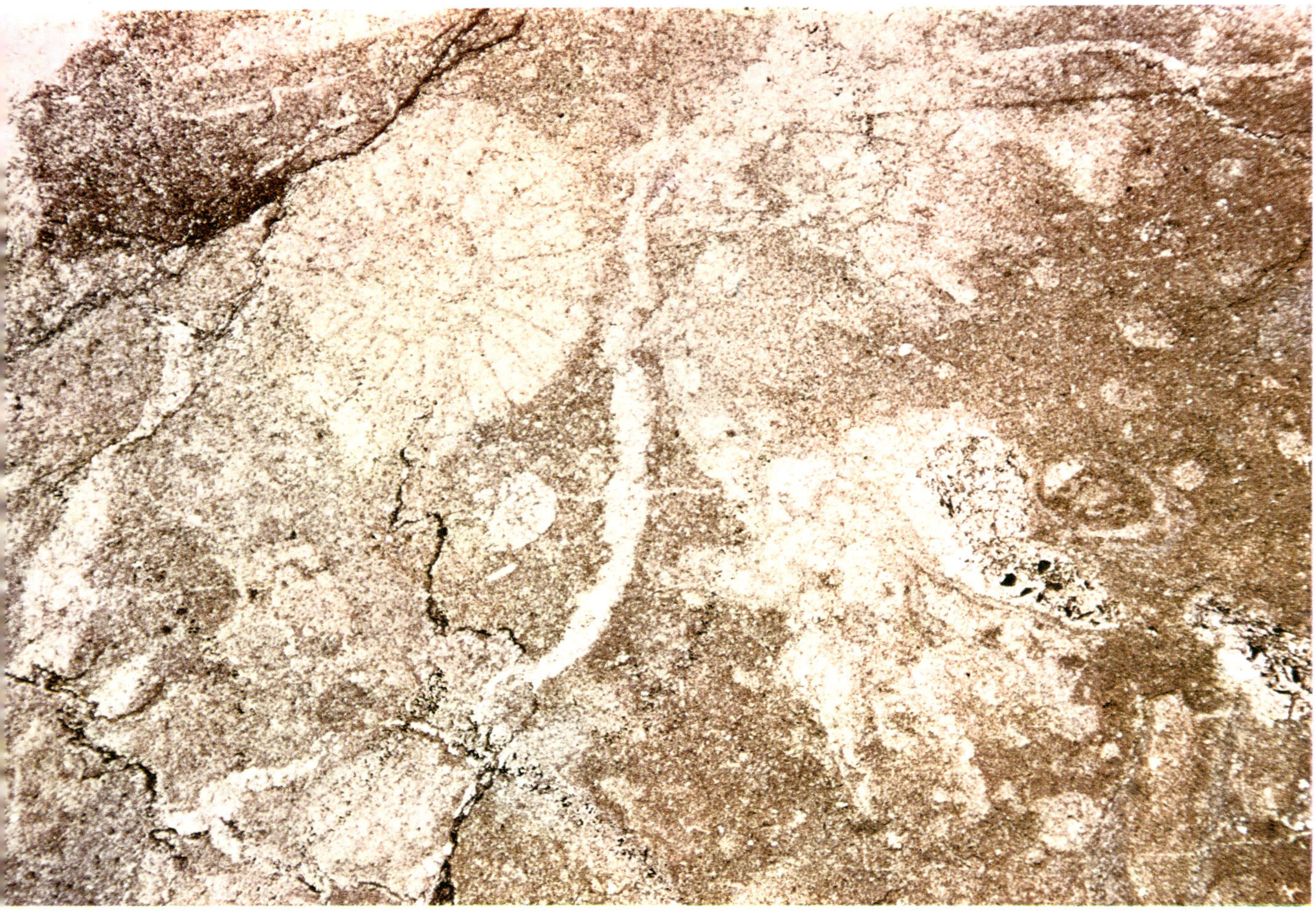

*Left: A thin section of limestone with archeocyatha of the Middle Cambrian. Inglesiente, Sardinia (×13). To the top left of the picture, a section shows the central cavity and the two concentric walls connected by radial septa*

sponges' belong; Hyalospongea, or siliceous sponges; and Calcispongea, the calcareous ones.

### Demospongea

The demospongea are the sponges that are most widely distributed in Nature today. The majority of them, however, disintegrate so quickly and completely after death that it is easy to understand why they are extremely rare in the fossil state. The demospongea are in fact mostly known through the spicules that are widely found in the sediments, or through the effects that they produce on other organisms. The finding in the very oldest layers of their characteristic tetraxial and monaxial spicules prove that they had enormously wide distribution, particularly in the Triassic and Jurassic periods. Some demospongea contain monaxial or tetraxial siliceous spicules.

Among these is the genus *Cliona*, a perforated sponge identified in fossil form by the holes that it makes in the shells of molluscs. It is very widespread in European Tertiary layers. Another common group, easily recognizable by virtue of the presence of siliceous spicules that secondary mineral deposits (desmide) have joined into a solid network, is that of the lithistides, dating back to the Cambrian period. There are some complete specimens of these in the fossil state in which the original form is visible.

### Hyalospongea

The hyalospongea are the siliceous sponges; their skeleton is composed of six-rayed spicules which in most cases are joined together to form an extremely light and transparent framework, from which their common name, 'glass sponge' is derived.

The body, shaped like a cup or cylinder set on the end of a long stalk, is extremely fragile, and as a result is seldom found intact in sediments; the typical spicules are more common, however, being formed in many rocks, from the Cambrian to the present time.

The present representatives of this class live chiefly in tropical seas, at an average depth of between 650 and 600 feet, but the deepwater forms are also found, dredged at a depth of 13,000 feet and, exceptionally, at 20,000 feet. However, many Cretaceous hyalospongea have been found in shallow sea sediments, at times directly in coastal locations, and it is thus logical to assume that within the last 60 million years they have migrated towards less shallow waters.

### Calcispongea

These sponges possess a very variable body form which is supported by isolated calcareous spicules (for the most part triaxial), and they live in isolation or joined in colony complexes in shallow sea waters. Their spicules are to be found in the rocks of the Cambrian period and later, but complete examples are not infrequent, among them the genus *Raphidonema* with its characteristic cup shape, which lived from the Triassic to the Cretaceous period.

# The Archeocyatha

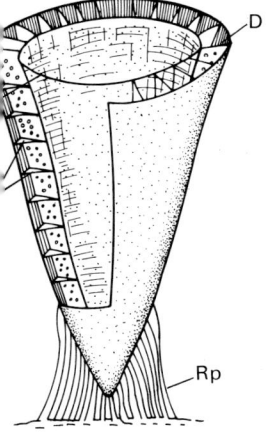

*Structure of the archeocyatha: Ew. External wall; Iw. Internal wall; Rp. Radicular process; T. Tabula; D. Vertical diaphragm*

The Cambrian sedimentary rocks of Southern Sardinia, Normandy, Sierra Morena, North America, the Sahara, Morocco, China, Siberia and Australia were formed in the sea 500 million years ago, and contain fossil organisms whose position in the whole palaeontological system has been under discussion for many years. Because of their vertical radial diaphragms, the fossils have been attributed from time to time to the coelenterata; others have ascribed them to the sponges, because of their porous walls and the presence of a central cavity; while another theory is that they arose from protozoa and the calcareous algae. Today these doubtful fossils are considered a phylum in their own right, and are called the archeocyathids. They are regarded as separate from the coelenterata and from the

sponges, having become extinct at the end of the Cambrian period without leaving any descendants.

The archeocyathids are conical organisms, of a height varying from 3 to 10 centimetres (1 to 4 inches). They consist of two concentric porous walls joined by vertical radial diaphragms, with horizontal tabulae that divide the area between the two walls into small, almost cube-shaped chambers. A large cavity, present at the centre, is similar to that observed in the sponges. The archeocyatha lived attached to the sea bottom by means of root systems from the apex of the cone, and probably inhabited shallow water near to the coastline. They had great importance as constructors of reefs, a role that was later assumed by the coelenterata.

# The Coelenterates

Thus the multicellular animals, or metazoa, developed in the Middle Archaeozoic period, a geological era in which their presence is documented by the discovery of a few fossils. Palaeontology has furnished the proof of their existence in that remote epoch, but cannot indicate the very earliest phase of their evolution because of the scarcity of remains, due on the one hand to the geological events that obliterated all traces of them, and on the other to the fact that such delicate organisms – largely devoid of hard parts – become fossilized only with difficulty.

On a rather more advanced level than the sponges are the coelenterata cnida, which have a primitive digestive tract. Some of their cells are differentiated to perform special functions, and there is a stratum of intermediate cells (mesoglea) between the external stratum (epidermis) and the internal stratum (gastrodermis). These are the first animals with defined tissues.

The coelenterata are well known because they are widely distributed in all the seas; few people can say that they have never seen, at least once, the slender branches of red coral or the clammy jellyfish. But how many would suppose that these organisms, so different from one another, could be classified in the same group? The fact is that in Nature the coelenterata are found in two different forms: one fixed or polypous, the other free or a jellyfish medusa. The polyp has the appearance of a cylindrical vessel, with a mouth aperture located at the top and surrounded by a corona of tentacles which adhere to the sea bottom by means of a disc. Anything entering the mouth aperture reaches a sac-like gastric cavity. The jellyfish, on the other hand, is shaped like an opened umbrella, and is continued at the sides in numerous tentacles. From the concave lower part (sub-umbrella) extends a formation called a 'shank', which carries the mouth. The body of the jellyfish contains up to 96.5 per cent water, and its tissues are more tightly packed than those of a polyp. The two forms described are characteristic of one group or another of coelenterata, or, more typically, they alternate within the same species.

It is believed that the oldest types of coelenterata were jellyfish, because in the older fossil-bearing strata of the Earth's crust impressions are preserved which are thought to be those of the umbrellas of these organisms. The polyp, on the other hand, generally has a very strong calcareous skeleton, and is entirely unknown in that remote era; it begins to appear in the seas of the Silurian period. Since that period the coelenterata often joined up in colonies of individuals and, because of their skeletons, they played a vital role in the formation of the sedimentary rocks.

If the oldest medusae are of greater interest to the scientist who is unravelling the evolution of coelenterates, the polyps, and in particular their skeletons, furnish collectors with an inexhaustible mine of finds from every geological era. They are often marvellously preserved and allow us to settle important palaeogeographic and palaeontological questions.

Because of the constructive role played by the coelenterates, they have been very useful in enabling scientists to obtain a picture of the sea bottom as it was millions of years ago. The powerful coral formations that they built at the start of the Silurian period indicate clearly the shifting of the seas at that time, and modern study of coral reefs shows the same kind of shifting pattern, which seems to have continued for more than 400 million years.

For coral formation to occur, organisms contributing to it must live in conditions suited to rapid development. The coelenterata are most delicate organisms that need very precise

*Right:* Astrangia lineata, *a colony of hexacorals. Miocene; Virginia Beach, USA. Diameter of the calyces, 0.4 cm*

*Below: A reconstruction of two of the oldest known coelenterate medusoids:*
*A. Dickinsonia, a dipleurozoite of the Australian Lower Cambrian;*
*B. Brooksella, a protomedusa from Algonquin Park in Canada. This has been partly sectioned to show both the external shape and the probable internal anatomy*

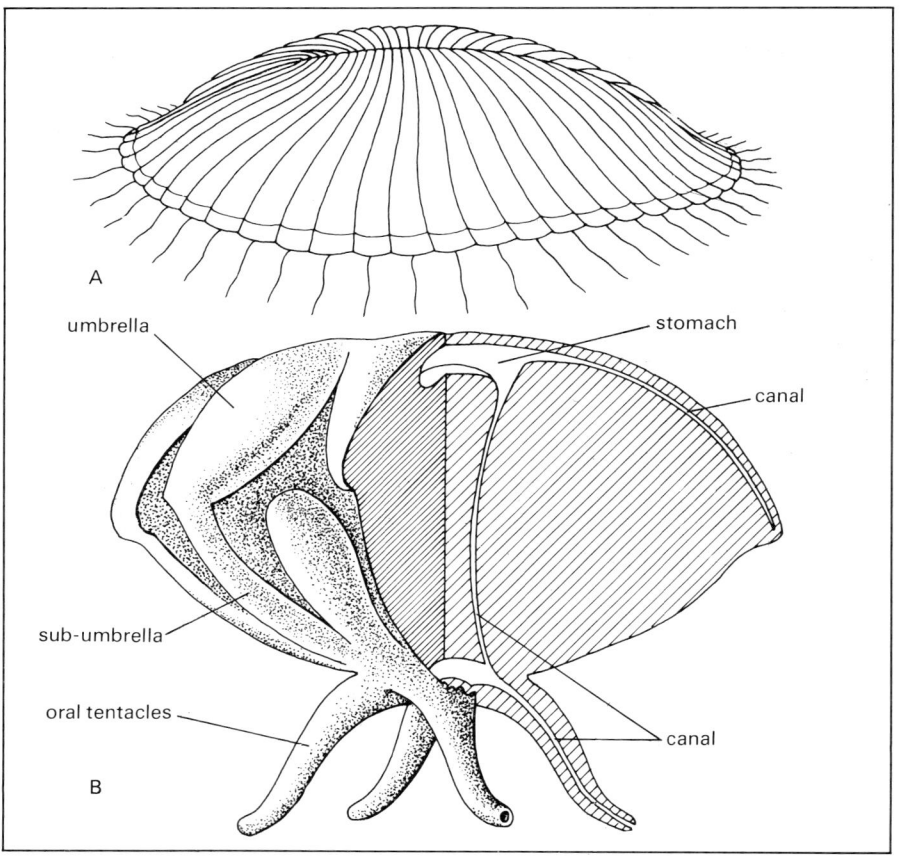

environmental conditions, without which death is inevitable. Consequently the manner of construction of present-day coral formations has been the subject of a great deal of research, and it has been established that the polyp of the constructing coelenterata requires the clearest of waters, at a temperature not less than 20°C, with a good light and thus no deeper than 170 feet, so that today their distribution has become limited to the area between the latitudes 27°S and 30°N, and only where there is no association with the cold marine currents. This rules out the west coast of Africa, which is washed by the Benguela current, and that of South America, where the Peruvian current passes.

The limit of depth to 170 feet appears to be in distinct contrast to the enormous thicknesses observed in both fossil and present-day coral constructions, but it has been observed that the part of the reef below the 170-foot limit is composed of skeletons of organisms that are now extinct. The problem arises of how it was possible for these thicknesses to be attained; it becomes clear that residues at greater depths indicate a progressive lowering of the sea bottom, while those lying above the upper limit of the tides indicate a rising of the bottom. Thus the thickness of the coral structures can be explained by a progressive lowering of the sea bottom at a speed equal to that of the growth of the upper part, this upper growth being due to the constructing coelenterata, which always live at the same depth.

There are three fundamental types of coral construction to be found in the seas at the present time: the coastal reefs which encircle the coasts of land masses; the coral barriers, lying parallel to the coast but separated from it by an arm of the sea; and the circular atolls enclosing a lagoon. The most majestic of coral constructions is the Great Barrier Reef, an imposing formation that runs parallel to the east coast of Australia for over 1200 miles at an average distance of about 65 miles from the mainland.

From the Silurian period onwards, numerous coral constructions were forming in virtually every one of the Earth's seas. Largely responsible for this construction, as we have said, were the large colonies of coelenterata, principally coral polyps, that lived on the sea bed. Cells situated on the constructing polyps extracted calcium carbonate from the sea water, and over a period of time this was deposited as limestone on the outside of the organisms. This hard skeleton then gradually enveloped the animals, which continued to grow upwards, new polyps budding from the top to form branching coral. Slowly, by this continuous repetition, the great coral reefs came to be formed.

These polyps, then, are found today in the fossil state, transformed into hard masses of coralliferous rock, within the stratified calcareous or dolomitic layers, and it is almost always possible to discover within these reef limestones the remains of the organisms that built them. Devonian coralline reefs, constructed 300 million years ago, are found in Italy, England, Germany, Canada, Australia and the United States. In the Triassic seas, which 170 million years ago occupied the north-eastern part of the Italian peninsula, the Dolomites were slowly constructed as a great coral barrier packed with

| Phylum | Class | Subclass | Order | Age |
|--------|-------|----------|-------|-----|
| Coelenterata | Protomedusae | | Brooksellida | Precambrian-Ordovician |
| | Dipleurozoa | | Dickinsoniida | Lower Cambrian |
| | Scyphozoa | Scyphomedusae<br>Conulata | | Cambrian-Recent<br>Cambrian-Triassic |
| | Hydrozoa | | Trachylina<br>Hydroida<br>Milleporina<br>Stylasterina<br>Spongiomorphida<br>Stromatoporoidea<br>Siphonophorida | Cambrian-Recent<br>Cambrian-Recent<br>Cretaceous-Recent<br>Cretaceous-Recent<br>Triassic-Jurassic<br>Cambrian-Cretaceous<br>Ordovician-Recent |
| | Anthozoa | Ceriantipatheria<br>Octocorallia<br>Zoantharia | | Miocene-Recent<br>Silurian-Recent<br>Ordovician-Recent |
| | | | Tabulata | Ordovician-Eocene |

remains of fungi, astreae and algae in calcareous form. They have attained an enormous height, thanks to the progressive lowering of the sea bottom that continued throughout the period. Reefs also abound in the Jurassic terrains of the Alps, the Apennines, in Switzerland and in many other sites inside and outside Europe, at higher latitudes than those at which they are at present found. In the Tertiary period the reefs vanished in Central Europe (where the corals remain only in isolated forms), while in the Mediterranean region others were constructed during the Eocene period; the best-known examples are in the neighbourhood of Verona and Vicenza, and indicate a climate much milder than at present.

As time brought with it a general cooling in climate, the reef formations were displaced progressively further south, finally reaching their present limits at the end of the Quaternary period.

Turning finally to systematics, which is of more interest to the fossil enthusiast, the abundance and the variety of coelenterata already to be found in the Lower Cambrian period indicates a very long evolutionary history for this phylum. From those remote times this history falls into three main periods, each corresponding to different classes; to these there have recently been added two new ones, which have very

ancient representatives. The coelenterata are thus divided into five classes, of which the first two, Protomedusae and Dipleurozoa, are exclusively fossils. The remaining three classes, Scyphozoa, Hydrozoa and Anthozoa, are still well represented in the seas of today.

### Protomedusae

The classification protomedusae covers certain primitive medusae whose impressions have been found in the Cambrian rocks of British Columbia, in the Ordovician rocks of Sweden and France, and in Precambrian terrains in North America. They are the oldest known remains of the coelenterata and are grouped under the genus *Brooksella*. Impressions are known of the discoid umbrella of this type, without tentacles. An age of 700 to 800 million years is assigned to the oldest of these impressions, found in terrains ascribed to the Algonkian period.

### Dipleurozoa

The dipleurozoa are impressions of medusae in the form of an elliptical bell with a deep median incision and numerous radial segments, separated by thin furrows. Numerous thread-like tentacles are present on the edges. The organisms are of the genus *Dickinsonia*, discovered in

*Above:* Conularia *sp., a scyphozoon of the now extinct subclass Conulata. Devonian; Bundenbach, Germany (×2).* Left: *Stromatopora concentrica, an encrusting hydrozoon of the order Stromatoporidae. Middle Devonian; Cologne, Germany (slightly enlarged)*

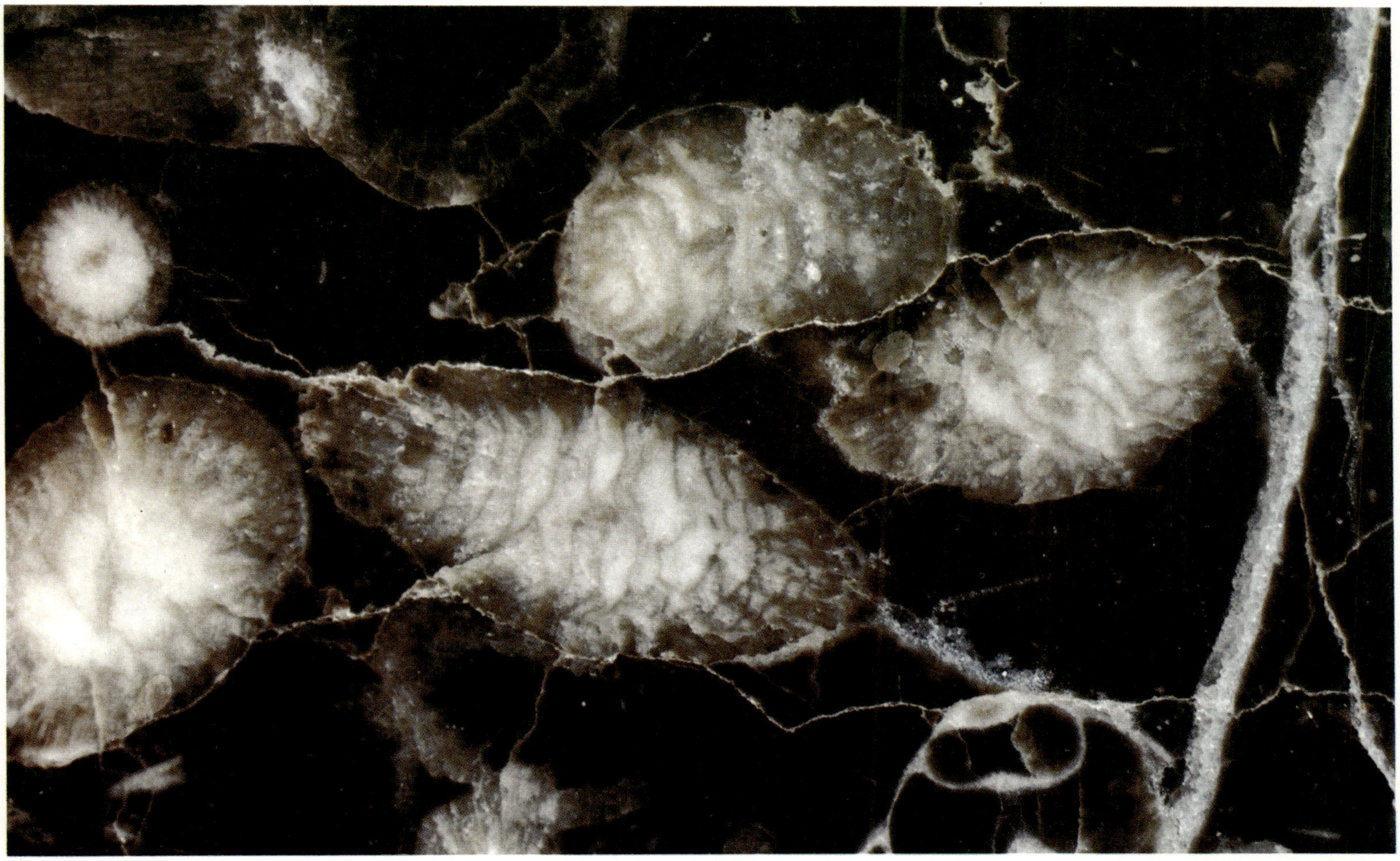

*A polished section of rock containing Lithostrotion martini, a tetracoral of the Carboniferous period. Belgium (×6)*

terrains of the Lower Cambrian period in South Australia, and they seem more highly evolved than the representatives of the preceding class. Like the protomedusae, they must have possessed a gelatinous body in the living state.

## Scyphozoa

These are organisms in which the medusa form prevails; the polyp, when it exists, represents only a transitional stage. The large medusae, up to two metres (six feet) in diameter, are grouped in this class. The scyphozoa are very rare in the fossil state and, lacking any hard parts, are represented exclusively by the impression left of the umbrella. They are grouped in the subcategory or class of Scyphomedusae, and have been found in very varied rocks from the Lower Cambrian to our own times.

The subclass Conulata comprises fossils that lived in seas from the Middle Cambrian period to the Lower Triassic, the period in which they became extinct. These were the coelenterata,

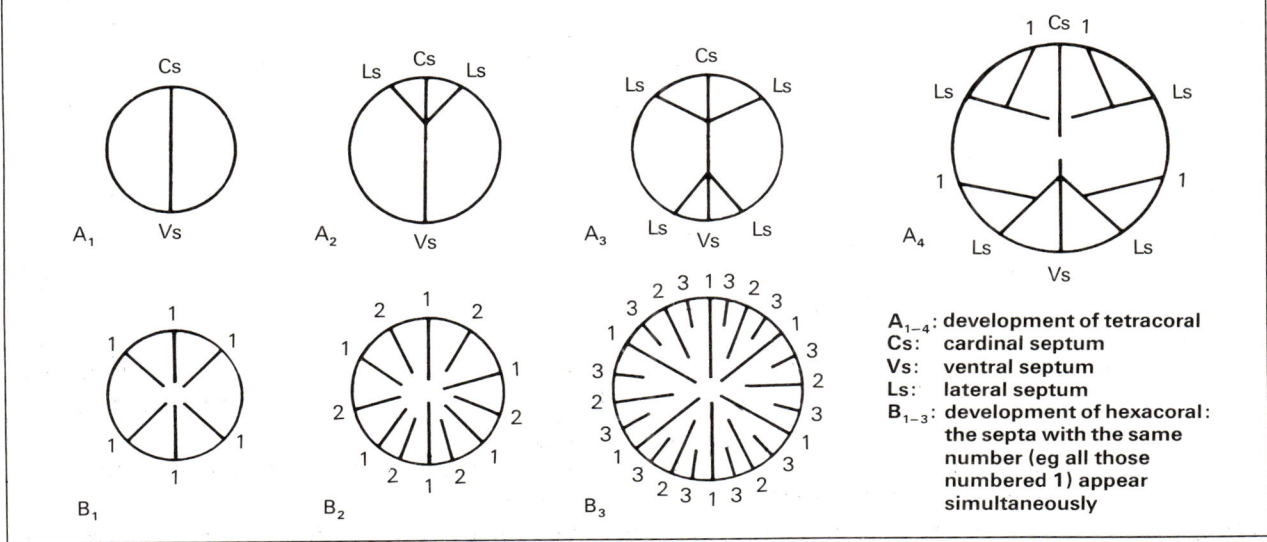

*Diagram showing the various stages of development of septa in the calyx of the zoantharia*

A₁₋₄: development of tetracoral
Cs: cardinal septum
Vs: ventral septum
Ls: lateral septum
B₁₋₃: development of hexacoral: the septa with the same number (eg all those numbered 1) appear simultaneously

which until recently were wrongly classified as either gastropods or 'worms'. They have an external chitino-phosphatic skeleton in the form of a pyramid varying in length between 4 and 10 centimetres (approximately 2 to 4 inches). The animal lived inside this shell, equipped with numerous tentacles that reached through an opening at the base of the pyramid, and which could be closed by means of movable plates.

It is thought that these organisms, which are not found today, lived fixed to the bottom of the sea by means of a disc applied to the apex of the shell during the juvenile state, and that in the adult stage they assumed a free-swimming jellyfish existence.

## Hydrozoa

The hydrozoa are coelenterata that are found as medusae and as polyps, either isolated or grouped in colonies. Forgetting the former, which are not often found in the fossil state, we will turn to the polyps, which are common in sedimentary rocks, and have played an important geological role in the development of coral formations.

The typical polypous hydrozoon is essentially a cylindrical tube consisting of a base, with which it attaches itself to objects, a slender upright stalk, and an elongated terminal polyp on which the mouth, surrounded by hollow tentacles, is to be found. The colonies are formed by shoots branching from the stalk and by buds on the body which give rise to other stalks. The stalks in their turn develop other terminal polyps and themselves reproduce. These colonies are generally strengthened and protected by a skeleton in the form of a chitinous tube. In other groups the polyps occupy holes that form in a massive calcareous skeleton. In many hydrozoa

*A section of the calyx of a siliceous tetracoral. Devonian of northern America ( ×4 )*

*Right: A magnificent example of* Calceola sandalina, *a tetracoral used as a guide fossil to the Devonian period. Germany.*

*Above:* Thecosmilia
trichotoma, *a hexacoral
of the Jurassic,
embedded in matrix.
The numerous radial
septa of the calyx are
clearly visible.
Württemberg. Diameter
4.5 cm (about 2 inches)*

the polyps within the colony are differentiated, and assume very different functions and appearances. In this way, there are large feeding polyps, small defensive polyps, and reproductive polyps.

The present class Hydrozoa divides the organisms into five orders: hydroids (Hydroida), milleporines (Milleporina), stylasterines (Stylas-terina), trachylines (Trachylina), and siphono-phorides (Siphonophorida), each of which has both fossil and living representatives. To these the palaeontologists add the spongiomorphid orders, which are exclusively fossil. We shall concern ourselves only with the most common types and those fossils that are most easy to find.

Isolated and colonial polyps as well as some medusae belong to the order of hydroids, which dates back to the Cambrian period. The polyps living today are to be found principally in coastal waters, in branched or encrusted colonies, and medusoids are to be found in open sea.

The genus *Ellipsactina* is characteristic. Animals of this genus are formed of oval nodules joined together by irregular pillars, each nodule being made up of thin layers of porous limestone. It is widespread in the Jurassic rocks in many parts of the world.

Constructing organisms which form calcareous ramifications or encrustations are classified within the order of Milliporina. Among these, the genus *Millepora*, with fossils dating from the Cretaceous period, is widely found today in the coral reefs, where it lives at a depth of about 30 metres (95 feet). It is formed by a simple calcareous crust, on the surface of which are numerous pores connecting with the interior

through vertical ducts. Within these are the digestive and defensive polyps.

Numerous coelenterata, now extinct, of particular importance during the Palaeozoic period in the construction of coral reefs, belong to the stromatoporoids. They were encrusting organisms possessing a massive calcareous skeleton, built up from numerous concentric thin laminae, or laminae. Each lamina, corresponding to a growth period of the animal, was in its turn formed by a certain number of smaller laminae, joined together by irregular pillars.

The outer surface of the colony is covered with holes that are connected to internal tubes inhabited by polyps, and 'astrorhizoids', star-shaped pores whose significance is not yet known.

The stromatoporoids, dating from the Cambrian period, are thought to have had a habitat similar to that of the millipores, which appeared in the Cretaceous period, after the disappearance of the stromatoporoids. In all probability they lived in warm shallow waters, contributing to the formation of neritic reefs, which were very widespread during the Silurian, Devonian and Jurassic periods.

### Anthozoa

The anthozoa are the largest group of coelenterata, in modern marine fauna as well as in fossil collections. They are organisms that live fixed to a submerged solid base, isolated or joined in colonies. Their principal features are a central gastric cavity divided by radial sectors and a ring of tentacles surrounding the mouth. The number of these tentacles varies in the different groups. There are six, or multiples of six in the zoantharia, and eight in the octocorals, which additionally often possess plumes. The importance of these organisms as constructors comes from their ability to secrete a robust skeleton, which is sometimes horny and preserved with difficulty, but in most cases calcareous, and thus easily fossilized. Vast accumulations of these and other organisms were responsible for the forma-

*Above left:*
*Hexagonaria sp.,*
*showing the typical*
*honeycomb-like colony*
*of these tetracorals*
*with polygonal calyces.*
*Devonian; Rockport,*
*Michigan ( × 1.5).*
*Above right:* Halysites
catenularia, *a tabulate*
*of the Silurian,*
*Gotland, Sweden. Note*
*the chain-like*
*disposition of the*
*tubules that make up*
*the colony ( × 2.5)*

tion of the reefs. A large number of anthozoa have a totally different structure, however, and lack this 'framework', and these are still the source of much speculation among geologists.

After a larval state, the young polyp moves freely about, and the animal starts to build a small cupola, known as a prototheca. After this, the walls of the calyx and the internal sectors are formed by calcification. The skeleton develops on the outside, and is made up of an external wall formed by the junction of the ends of the inflated sectors (false wall) or by calcification that does not affect the sectors themselves (true wall); in both cases a rough calcareous crust (the epitheca) envelops the wall. The cavity of the calyx of a coelenterate is thus divided into a number of lobes by radial sectors. These do not however meet at the centre, but leave a space occupied by a vertical pillar, or columella. This very complex structure, which has in fact been simplified in the above description, is to be found in all the anthozoa, in both the isolated forms (simple polyps) with a conical calyx, and in the colonies, which are made up by a variable number of calyces joined together (colony polypierites). The colony polypierites may be different in appearance, but are notable for their beauty and variety. They have branched forms, in which

the arms are free, and larger forms in which the various calyces are joined together by a common tissue (coenenchyma). In a very few cases the polyps join to form a polygon or – exceptionally – some of them lose their individuality and come together to form a labyrinth.

Fossilized anthozoa from the Ordovician period are known and are divided into three subclasses – Ceriantipatharia, Octocorallia and Zoantharia – all with representatives still living. Of these subclasses, only the last two have a definite value in palaeontology because of the abundance in which their fossils are found.

The octocorals include the halcyonaria, colonial forms whose skeleton – when there is one – is in the form of a shaft or of calcareous spines joined by a horny substance. Most fossil finds of animals in this group are simply isolated spines in the sediments, since the cementing horny substance disintegrates after death, rarely allowing the preservation of the complete animal. Among the best-known forms is the red coral used in jewelry. This was already growing on the sea bottoms of the Cretaceous period, more than 60 million years ago.

The zoantharia include very varied organisms: those without a skeleton, like the well-known actiniae, or sea anemones, cannot be preserved

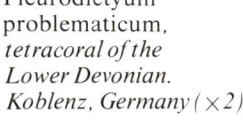

Pleurodictyum problematicum, *tetracoral of the Lower Devonian. Koblenz, Germany (×2)*

Heliotes *sp., tubular,
Silurian; Germany*

as fossils, while others which had a very durable calcareous skeleton remain, and can be seen in all their elegant detail, many millions of years later. These, the madrepores, still play the same part today in the construction of coral reefs as they did in the past.

The zoantharia are classified by palaeontologists into tetracorals and hexacorals according to the arrangement of the sectors in the calyx. The first group is exclusive to the Palaeozoic period and became extinct at the end of the Permian, to be replaced in the following period, the Triassic, by the hexacorals. These have survived without any great modifications for 170 million years, and are today a vigorous example of zoantharia.

As we have said, the names tetracoral and hexacoral refer simply to the grouping of the sectors in the calyx: groups of four in the tetracorals, and groups of six in the hexacorals. In the growth of a tetracoral, the first stage is the appearance of a diametrical septum. Later two lateral septa are produced, one on each side of the diametrical septum and leaning against it, and these are followed by two septa at the opposite side to the diametrical septum. Soon the diametrical septum splits to form cardinal and ventral septa, while the other septa team up in groups of four. The arrangement of the groups, two between the laterals and the cardinal on one side and two outside the laterals on the other side, results in the bilateral symmetry that is a feature of the adult organism.

Hexacorals, on the other hand, display radial symmetry and this is established in the earliest stages of growth of the calyx. The septa of this type of coral appear in groups of six simultaneously and the successors in each septum are formed from the preceding ones.

Although the distinction between tetracorals and hexacorals sounds straightforward, most people find it difficult to establish which of the two groups a fossil belongs to without the advice of a specialist, who has at his disposal a microscope and thin or transparent sections of the fossil. The external form of the representatives of the two groups is so similar as to be deceptive on most occasions.

Finally, it is necessary to attribute the tabulates (Tabulata) to a separate group of almost exclusively Palaeozoic coelenterata, which, because of their structure, do not seem to belong in any of the preceding categories. They are in fact colonial organisms, now extinct, whose skeleton is formed by calyces in the form of small tubes laced by horizontal septa.

# The Bryozoa

Ascending the evolutionary path towards constantly more complex animals we now reach a group of organisms little known to scientists, for whom they constitute a virtually insoluble conundrum. These are the bryozoa, or sea-mats, invertebrates whose history is obscure: some people classify them into a phylum of their own, while others divide them into two independent groups. Up to the present time nobody has discovered any affinity between these and any other animal groups, with the possible exception of the brachiopods – themselves a zoological enigma – which seem to be the closest relatives of the bryozoa.

The bryozoa were once included among the 'zoophytes' (literally, animal-plants), because they were considered to be intermediate organisms between the plants and the animals, their external form being that of a vegetable. Subsequently, the discovery that their appearance was due to a combination of several small details induced some scientists to put them close to the coelenterata, and following this it was observed that each individual of the colony

had a complete alimentary canal, with a mouth aperture, stomach, and anal aperture. This observation again raised the question of classification, resolved by some scientists through the institution of a group of the vermitides, including both bryozoa and brachiopods. For present purposes, however, the bryozoa will be regarded as a single phylum.

Although the bryozoa are very common animals in the fossil state, their collection and study are quite difficult because of their small size. For accurate classification they must be observed with microscopes of high magnification, frequently in thin sections, while even their collection in most cases requires the techniques commonly employed for microfossils. However, fragments of colonies visible to the naked eye are not rare, and the collector can take these home and establish that they are bryozoa; he will not be able to apply a more detailed classification unless he is a specialist in the field. The very general classification – simply identifying the animals as bryozoa – is possible through a knowledge of their structure when in the fossilized

*Encrustation of Cellepora sp., one of the bryozoa of the order Cheilostomata, on the shell of a Pliocene gastropod. Britain (×4 approx)*

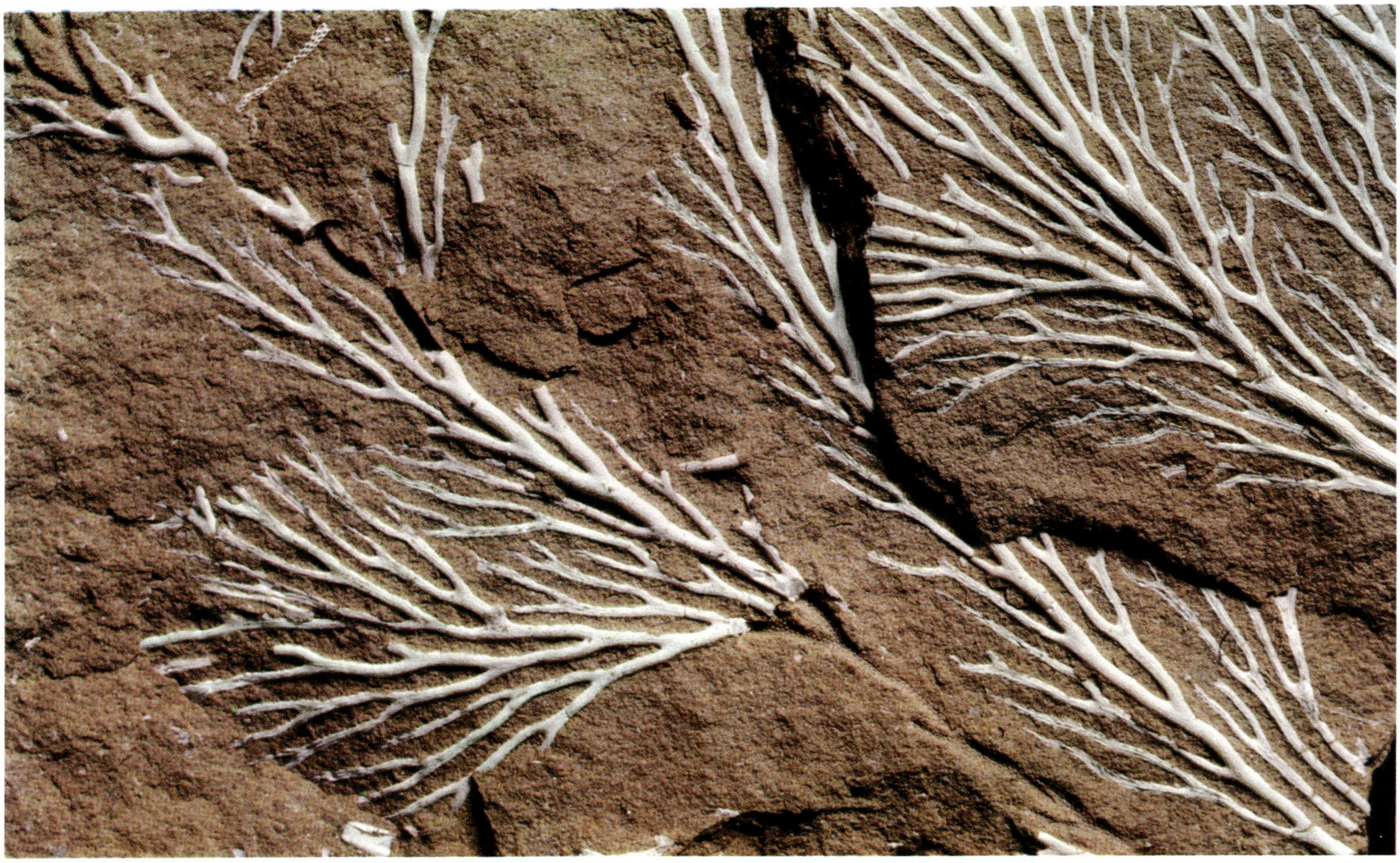

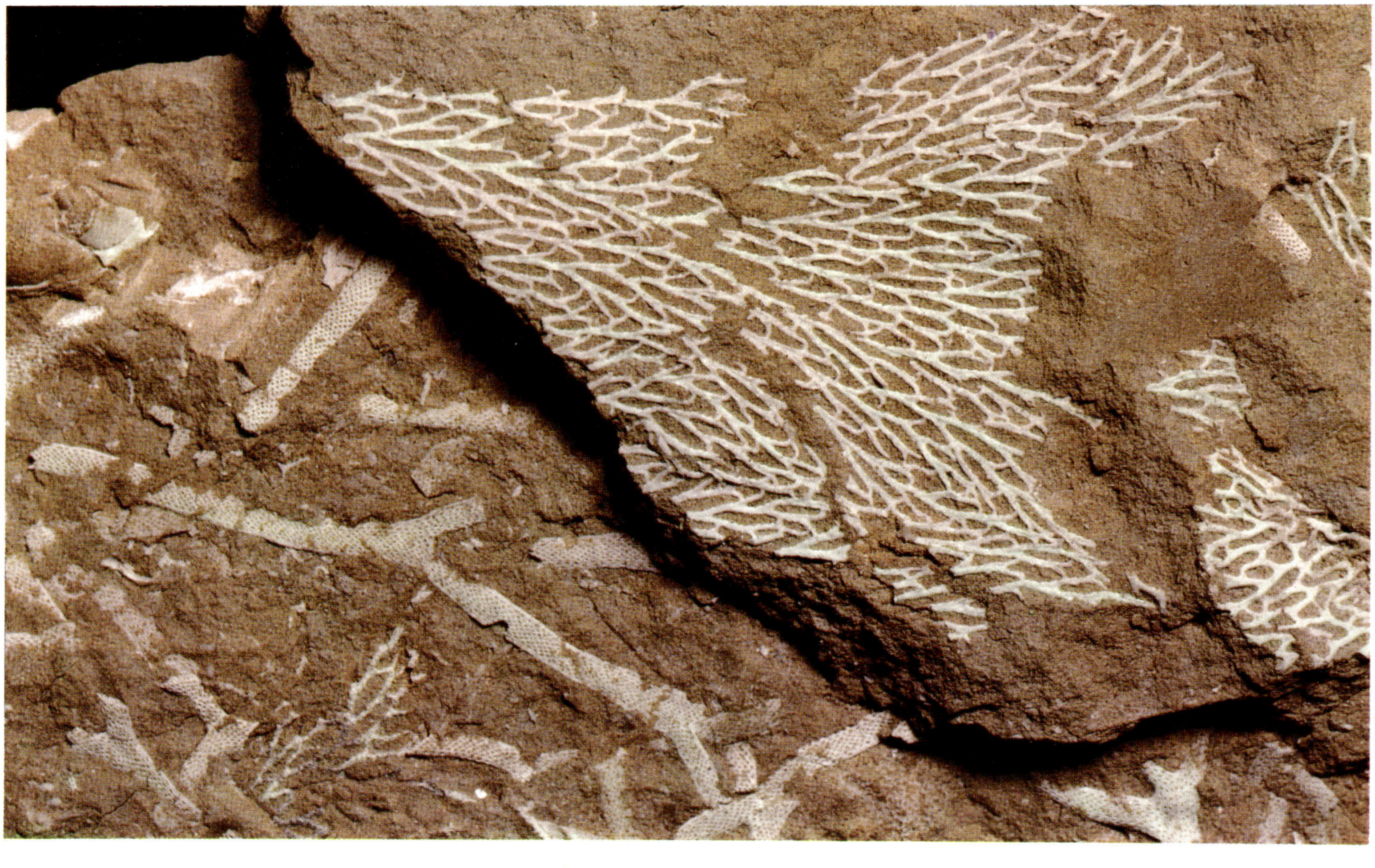

| Phylum | Subphylum | Class | Order | Age |
|--------|-----------|-------|-------|-----|
| Bryozoa | Entoprocta | | | Recent |
| | Ectoprocta | Phylactolaemata | | Cretaceous-Recent |
| | | Gymnolaemata | Ctenostomata<br>Cyclostomata<br>Trepostomata<br>Cryptostomata<br>Cheilostomata | Ordovician-Recent<br>Ordovician-Recent<br>Ordovician-Triassic<br>Ordovician-Permian<br>Jurassic-Recent |

state, without going into any great detail.

The bryozoa are colonial invertebrates that today live mainly in clear and moving waters of variable depths; some rarer representatives are adapted to life in fresh water. From their first appearance in the Ordovician period, about 430 million years ago, the bryozoa have played an important part in marine animal life, and today they are found in the fossil state, at times in great profusion. These fossils exist mainly in argillaceous and calcareous sediments, and their nature seems to indicate deposition in shallow waters.

Although the appearance of individuals in the various genera may be fairly constant, whole colonies of bryozoa differ greatly in size and shape. As a rule they grow on shells, pebbles or other hard bodies, forming very delicate encrustations whose structure may be lamellar, hemispherical, irregular or tufty. Considering the great abundance of these organisms, it is not surprising that fragments are frequently found on the shores after very high tides. Sometimes, in certain conditions, these fragments are so numerous that they accumulate to form large expanses of 'bryozoan sands'. This phenomenon is known to have occurred in some geological periods, because of the finding of genuine 'bryozoan rocks' which are nothing other than fossilized sands.

Thus the fossil bryozoa are not altogether rare within sedimentary rocks. Most notable are some English and North American deposits which date back to the Silurian period and in which fragments of the organisms completely cover the surface of the rocks. In Europe they are abundant in the Eocene and Oligocene layers of Vicentino, Italy, and Germany, where in certain lake regions organisms of the Permian period have formed prominent reefs. There the bryozoa, in association with brachiopods, molluscs and other organisms, rise up as islands in the middle of the lakes.

A colony of bryozoa is an assembly of small polyps, covered with a chitinous or calcareous skeleton. It is formed by the freely swimming larvae of one individual, which is fixed to a solid object and produces all the other organisms of the colony through budding. Each individual organism, less than a millimetre long, is called a zooid. It has a sac-like body with a ring of tentacles (the lophophore) on the upper part. At the centre of this is the mouth. The external wall of the sac covers a chitinous or calcareous sheath, with an aperture in which the animal lives; the sheath (carapace) is the only part that is fossilized. The complete colony (zoarium) is thus composed of a large number of zooids that inhabit small carapaces, mostly in the form of polygonal tubes.

On the outer side of some carapaces are special structures known as avicularia (because of their resemblance to a bird's head), which protect the individuals from predators. On the outer side there may also be long vibratile hairs, called vibracula, which seem to have the function of preventing larvae or noxious materials from entering the sheath.

The phylum Bryozoa is divided by palaeontologists into two subphyla: Endoprocta, in which the anus opens inside the circle of tentacles; and Ectoprocta, in which the aperture is outside the lophophore. The first type, not having any hard parts, is not known in the fossil state. The Ectoprocta, on the other hand, are very abundant in the rocks of all eras, and are subdivided into two groups; the Phylactolaemata, which are the only bryozoa adapted to life in fresh water and are known fossilized from the Cretaceous period;

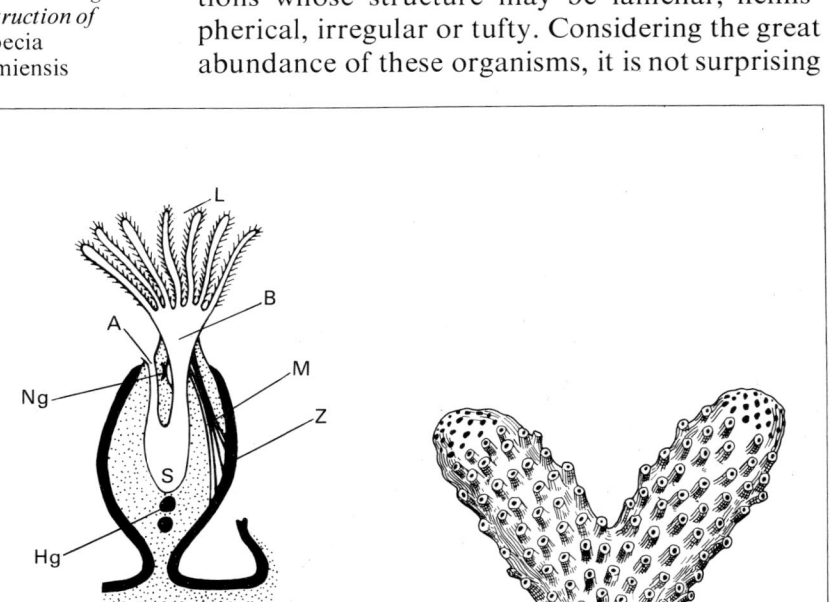

and the Gymnolaemata, known from the Ordovician period, which are grouped into five orders. These orders are the Ctenostomata, Cyclostomata, Trepostomata, Cryptostomata, and Cheilostomata.

The Ctenostomata existed in the Ordovician period and still live in marine and estuarine water. They are boring bryozoa, with horny or gelatinous carapaces that are not well adapted to preservation, and are known to palaeontology mainly as a result of the excavations that they have left on pebbles, on rocks and on the shells of molluscs and brachiopods.

In the order Cyclostomata, the grouped colonies are composed of tubular calcareous carapaces. The free ends of the protruding carapaces have a large aperture with no operculum. They are very old and abundant bryozoa, found for the first time in Ordovician strata, and very widespread during the whole Palaeozoic era. They reached their greatest development in the Jurassic period, began to diminish during the Tertiary period, and today only few are found.

The Trepostomata form colonies of elongated cylindrical or prismatic tubes, sometimes joined into compact shapes, and in other cases in very delicate ramified forms. The oldest are from the Ordovician period, a time when they had their

Fistulipora carbonaria, an encrusting bryozoa of the order Cyclostomata. Upper Carboniferous; Eastland, Texas (×2.5)

Fenestella bohemica, a bryozoa of the order Cryptostomata. Devonian; Koněprusy, Czechoslovakia

maximum development. After that began the slow decline that brought them virtually to extinction at the end of the Palaeozoic era. According to recent data it seems likely that they were present, if only in greatly reduced numbers, during the Triassic period.

The Cryptostomata build large colonies that have the appearance of nets or laminae, formed from elongated individuals with a square or hexagonal section. They are exclusively Palaeozoic, having lived during the Ordovician and Permian periods and having been responsible for important reef formations.

Finally, the representatives of the order Cheilostomata are among the most 'modern' of the bryozoa, since they only appeared in the Middle Jurassic period and are now easily the most common in present-day waters. They form ramified or compact colonies, made of oval or elliptical individuals side by side. They have a small nonterminal aperture with a movable operculum.

# The Brachiopods

*This* Terebratula *sp. of the Quaternary, from Calabria, has the classic 'oil lamp' shape common to many brachiopods, caused by the difference in development of the two valves and the large perforated umbo*

The brachiopods, or lamp-shells, are a group of small marine invertebrates that live on the sea bottom, and which were one time thought to be closely related to worms, along with the bryozoa, which have a very similar general structure. The brachiopods have, however, evolved further and abandoned the formation of colonies. They still exist as individuals in close communities, and as a consequence have developed in size. They also possess a bivalve calcareous or chitino-phos-phatic shell which is easily fossilized, and is the part of the organism that principally interests the palaeontologist, who often bases his classifications on its various characteristics.

The shell of a brachiopod is symmetrical. It consists of a large ventral valve at the rear end of which is a curved umbo, often perforated, and a smaller dorsal valve. The similarity of the shell to that of other bivalves such as the clams often creates problems for the amateur, who may be

uncertain how to classify his find, but three main features are characteristic of the brachiopods, and permit identification. Unlike those of most bivalves, the two valves of the brachiopods are different from each other, lying above and below the soft body. The umbo only occurs on the ventral valve, and in most species has a hole in it which serves as a passage for the pedicle (stalk) by which the animal attaches itself to the sea bottom.

Without giving a detailed description of the soft parts of these strange animals, we should mention that the valves are covered inside by two membranes which form a sheath or mantle, enclosing a cavity in which is located the visceral sac containing the various soft organs, and the lophophore, an organ similar to that seen in the bryozoa, which is formed by two coiled fringed strips or brachia located at the two sides of the mouth. By means of whip-like movement of the brachia a flow of water is produced, from which

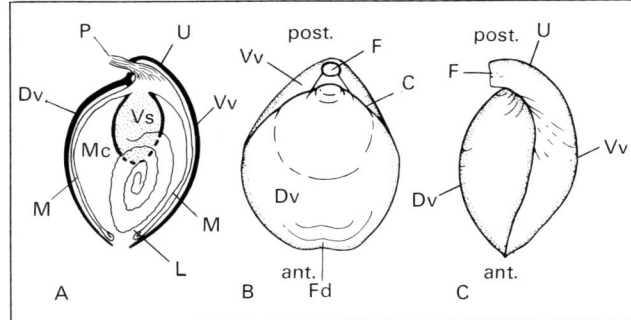

*Morphology of a typical brachiopod (A. Internal anatomy; B. Dorsal view; C. Lateral view); Ant. Anterior margin; Post. Posterior margin; Mc. Mantle cavity; F. Foramen; L. Lophophore; C. Cardinal line; M. Mantle; P. Peduncle; U. Umbo; Fd. Fold; Vs. Visceral sac; Dv. Dorsal valve; Vv. Ventral valve*

*Below left: An inarticulate brachiopod, Discina calymene. Triassic; Pedraces, Bolzano, Italy. Below right: Lingula cuneata, a representative species of the most long-lived group of inarticulate brachiopods, which has remained almost unchanged for some 400 million years. Lower Silurian; Medina, New York (×1.5)*

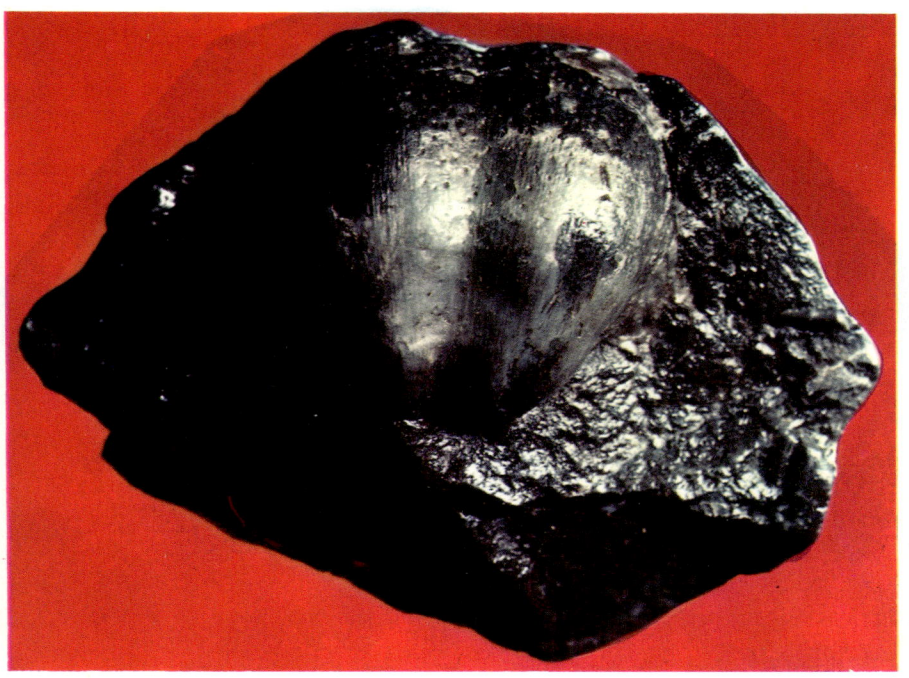

seen in brachiopods deformed as a result of being fixed in particular surroundings. In addition, the shell may be thin in the deep-water forms and thick in the coastal or reef forms, while its surface may have a smooth finish or be ornamented by ribs, stripes or spines.

In some brachiopods, the two valves of the shell are joined to each other by means of a simple hinge, which enables the shell to close and open. The first subdivision of the brachiopods into the two classes called *inarticulates* and *articulates* is, in fact, based upon the presence or absence of this fundamental apparatus. In the first class the hinge is lacking and the two valves are joined exclusively by strong contractile muscles, while in the articulates the hinge is formed of two teeth symmetrically disposed below the umbo at each side on the ventral valve, and by two recesses into which the teeth fit, on the dorsal valve. In both the inarticulate and the

*Above:* Horridonia horridus, *an inarticulate brachiopod of the order Strophomenida. Permian; Kerborn, Germany (×1.5). Far right:* Tetractinella trigonella, *an articulate brachiopod and a guide fossil for alpine Triassic beds. Vicenza, Italy (×2)*

the animal derives food and oxygen. (Interestingly, the name brachiopod itself, which is derived from two Greek words meaning 'arm-foot', arose from the mistaken idea that the brachia were loco-motive organs.)

The lophophore is sometimes supported by a calcareous framework that varies in shape from species to species. This is known as the brachial apparatus or brachidium. It is often preserved in the fossil and may assist in its classification. Such an apparatus does not exist in the inarticu-lates. In the more evolved forms it takes the appearance of two small symmetrical apophyses, termed crura, which in the most evolved brachio-pods are elongated and rolled in the form of a spiral cone.

A large part of palaeontological classification is thus based upon the shell, the calcareous, the calcareous-horny and the simple horny body structures. They may be very different in shape: they can be oval, globular, hemispherical, plate-like, convexo-concave, or irregular – as may be

| Phylum | Class | Order | Age |
|---|---|---|---|
| Brachiopoda | Inarticulata | Lingulida<br>Acrotretida<br>Obolellida<br>Paterinida | Cambrian-Recent<br>Cambrian-Recent<br>Cambrian<br>Cambrian-Ordovician |
| | Articulata | Orthida<br>Strophomenida<br>Pentamerida<br>Rhynchonellida<br>Spiriferida<br>Terebratulida | Cambrian-Permian<br>Ordovician-Jurassic<br>Cambrian-Devonian<br>Ordovician-Recent<br>Ordovician-Jurassic<br>Devonian-Recent |

articulate brachiopods the movements of the two valves are controlled by divaricator and adductor muscles, which serve respectively to open and close the valves. There are also pedicle muscles that move the stalk by which the animal attaches itself to rocks, and these allow sideways movements of the shell. All these muscles leave a characteristic muscle impression on the inside of the shell, which can be seen clearly in many fossils and is sometimes used to distinguish a particular species or genus. In some very evolved forms the muscles are fixed on a special organ shaped like a spatula, called a spondylium.

The brachiopods are exclusively marine animals that may live attached to one another, or on submerged objects on the sea bed, covered in sediment, often in compact groups of individuals. The organism attaches itself by means of a pedicle which leaves the shell through a special hole or foramen in the umbo of the ventral valve. This varies in shape and size according to the species of the animal, and is therefore a useful distinguishing feature. Some fossils in fact had different methods of attaching themselves; some used the entire ventral valve and others a series of long spines that covered either the whole surface of the shell or only the rear edge.

The brachiopods are to be found today in all the seas, but principally in the warmer ones on the sea bed at depths varying from a few metres to some 200 metres (650 feet). Rare forms have however been dredged from depths of as much as 5,000 or 6,000 metres (16,000 to 17,500 feet), indicating the remarkable adaptability of the whole group. This adaptability is also found in the fossils, which are much more numerous than the present forms, and have been found in very different habitats. In this connection, it is interesting that the genus *Gemmellaroia*, which was living during the Permian period, about 200 million years ago, was adapted to a reef existence, and has a shell modified for the purpose. One valve is elongated, and conical in shape, and the second transformed into an operculum. As a result of these modifications the organism appears very similar to certain coelenterata and to the acephalous lamellibranchs that make up the reefs of the Cretaceous period.

At the present time, there are about 260 species of brachiopods, representing over 70 genera. Several species are found in American waters, and round the British Isles, but they are most common in the Caribbean, the south-west Pacific, the Mediterranean, and off the Atlantic coast of Africa. Brachiopods were very much more common in past epochs, when they are thought to have consisted of over 25,000 species,

*Prionorhynchia quinqueplicata, an articulate brachiopod of the Lower Jurassic. Gozzano, Novara, Italy. Note the zig-zag edge where the two valves meet. Size of the original, 3 cm (about 1 inch)*

constituting a considerable part of the marine fauna. Their remains are thus used by palaeontologists as fossil guides for many geological periods and provide an excellent means of correlating rocks of the same age over a wide area.

For example, an excellent guide fossil for the Palaeozoic terrains is the genus *Cyrtospirifer*, an articulate of the order Spiriferida, and particularly characteristic of the Devonian period. This exhibits a shell formed by convex valves, with an extremely elongated straight rear edge, ornamented by radial ridges. The genus *Productus* has also furnished a good number of guide species for various geological periods. Animals of this genus and other articulated brachiopods of the order Strophomenida were found widely distributed in the Silurian, Devonian and Permian periods. They possess a flat convex shell with a highly developed umbo on the ventral valve, which is inflated and elongated along the hinged

*Pygites diphyoides, an articulate brachiopod of the order Terebratulida, from the Upper Jurassic (×1.5). It is one of the most characteristic brachiopods, and is used as a guide fossil*

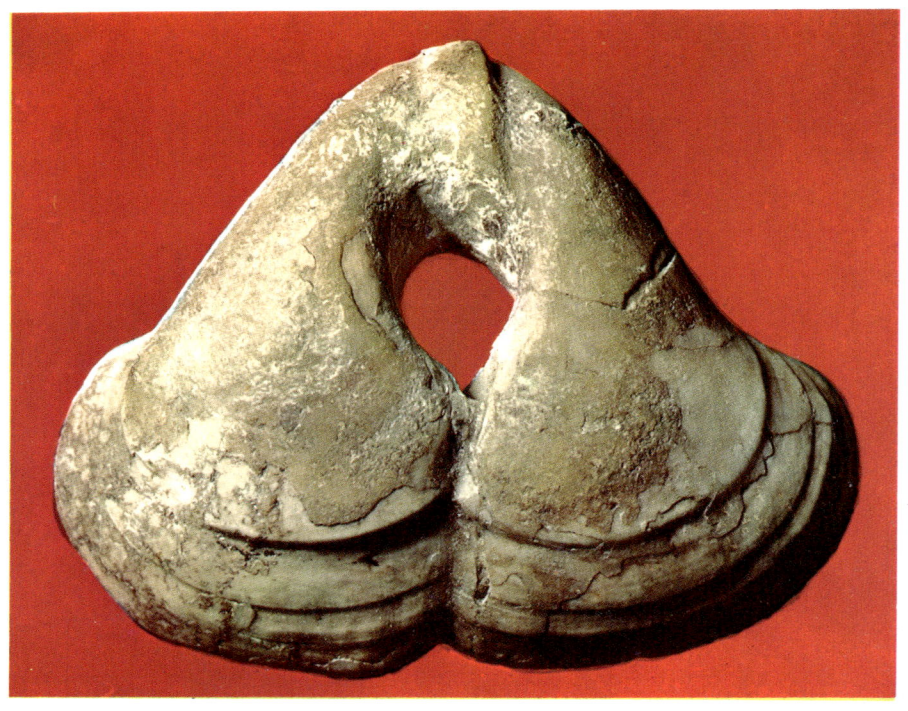

Paraspirifer *sp., an articulate brachiopod of the order Spiriferida, characteristic of the Devonian era. Ohio* (×3)

edges, where it forms two lateral auricles. The two valves are ornamented by radial and concentric ridges which give the appearance of a network. On the ventral valve there are very sophisticated spines with which the animal fixed itself to the bottom, instead of with a stalk. Among the most noteworthy species belonging to the order Productidina is *Gigantoproductus giganteus*, which at 30 centimetres (12 inches) across, is undoubtedly the largest brachiopod that has ever existed.

A genus adapted to reef life in the Middle Devonian period is *Uncites*, an articulate belonging to the order Spiriferida. They are unusual brachiopods, with a large biconvex oval shell, ornamented by concentric growth rings and radial ridges. Their ventral valve is much more developed than the dorsal, and bears an enormous curved and hooked umbo.

The brachiopods are one of the oldest groups of all the invertebrates. Lower Cambrian strata in Montana have yielded what are considered to be some of the oldest known brachiopods. They are shaped like an oval valve and are tentatively attributed to the genus *Linguella* of some 800 million years ago. In the Lower Cambrian, inarticulates were already well developed: some of the genera have been preserved unchanged to

this day, through more than 500 million years. In the same period the first articulates also appeared. These were to have a most impressive development, mainly in the Silurian period, when they numbered about 3,000 species. Throughout the Palaeozoic and Mesozoic eras the brachiopods were to flourish; their decline in fact only began in the Tertiary period, an era in which they diminished progressively until reaching roughly the restricted distribution that they have today.

Brachiopod remains are to be found in most marine sediments the world over, but the Lower Palaeozoic rocks of central Europe, some parts of Scandinavia, the Midlands of Britain and the North American continent as a whole are rich in brachiopod fossils. Whole faunal assemblages can be seen in Middle and Upper Palaeozoic strata of the Glass Mountains, Texas, and many species are also represented in the Lower Palaeozoic strata of New York, Indiana, Ohio, Oklahoma, Kentucky and Wisconsin. They are frequently used as guide fossils. Just one example is the genus *Tetractinella*, an articulate of the order Spiriferida, which has a small five-sided biconvex shell with four pronounced ridges on each valve radiating from the umbo. This genus is very common in Alpine Triassic strata, where the species *Tetractinella trigonella* is character-

41

*Left: A slab of brachiopods (Spiriferida) from the Devonian of Koblenz in Germany (×2)*

*Below:* Productus cora, *an articulate brachiopod of the order Strophomenida. Carboniferous; Austria (×3)*

istic of the middle part of that period. The curious *Pygope*, an articulate of the order of Terebratulida, is another typical fossil. This has a central hole formed by the rapid growth of the lateral parts of the front edge of the shell, welded on the median line. It is characteristic of the Jurassic and Cretaceous terrains and has furnished numerous guide species for the rocks of these periods.

The phylum of the brachiopods is thus, as we have said, divided on the basis of the absence or the presence of the hinge into the two classes Inarticulata and Articulata. Both of these include numerous orders, which we shall mention briefly.

**Inarticulata**

The inarticulates are brachiopods with a chitinous or, in some cases, calcareous shell without a hinge, in which the valves are held together by muscles. The stalk, or pedicle, when there is one, grows from the valves or through a hole in the ventral valve. There is no brachial apparatus. The inarticulates are the oldest, and possibly the most primitive brachiopods known. They range from the Cambrian period and may still be found living in some oceans today. They comprise four orders: Lingulida, Acrotretida, Obolellida, Paterinida.

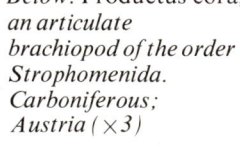

The representatives of the Lingulida are brachiopods with a biconvex or conical chitinous or – more rarely – calcareous shell, which is either smooth or ornamented, with light concentric growth rings and with beak-shaped umbos on both the valves. The pedicle comes out on the line of separation of the two valves at the back of the animal. They date back to the Lower Cambrian period and are still to be found in the seas today.

The best-known genus is *Lingula*, which has a thin elongated shell, equipped with a long pedicle that springs from the sharp rear edge. This genus, which was common in the Ordovician and Silurian periods, has been preserved virtually unchanged for over 400 million years.

The Acrotretida have a circular or conical shell, which is either chitinous or calcareous, and smooth or ornamented with more or less marked concentric rings. A peduncular aperture appears on the ventral valve near the edge in the less evolved forms, and at the centre in the more specialized. The order is known from the Lower Cambrian period and some representatives of it are extant.

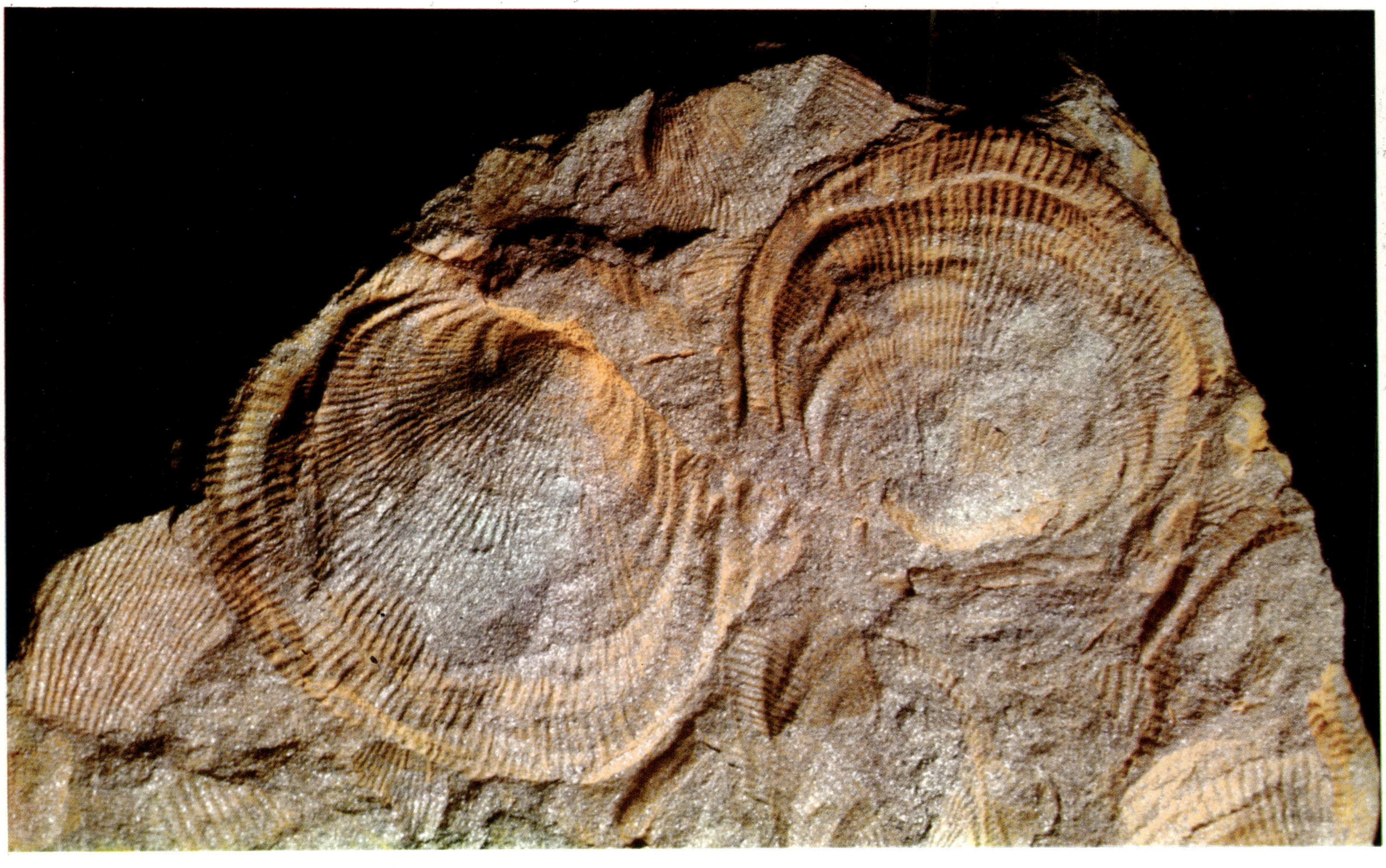

The order Obolellida comprises a few genera alive exclusively in the Cambrian period which started to become extinct around the middle of this period. They had a semicircular or oval biconvex calcareous shell, with a long pedicle emerging from the ventral valve, or a pedicle aperture that varied in position.

Paternida are one of the earliest brachiopod stocks, ranging from the Lower Cambrian to Middle Ordovician. The internal structure of these forms is unknown, but externally they vary in shape from oval to semicircular in general outline and are often faintly conical with numerous concentric lines or growth marks. The composition of the shell is chitino-phosphatic and some species have large indentations or superficial perforations on the shell outer layer. There is no pedicle or foramen in this group.

## Articulata

The articulates are brachiopods with a calcareous shell equipped with a hinge apparatus formed by teeth and recesses. The hinge is always present, although the shape varies, and the pedicle passes through an aperture that affects the ventral valve only. The articulates seem to be derived from the same ancestors as the inarticulates, and are differentiated from these last by the progressive modifications that justify their classification into two separate categories. In fact the differences between the oldest articulates and the inarticu-

*Right: A slab containing specimens of* Pugnoides *from the Lower Carboniferous period. Britain (×2.5). Below: Different types of Rhynchonellida in Jurassic rock from Britain*

lates are not nearly as obvious as those between the modern surviving representatives of the two classes.

The articulates are subdivided into six orders: Orthida, Strophomenida, Pentamerida, Rhynchonellida, Spiriferida and Terebratulida.

The representatives of the Orthida have a hemispherical or almost rectangular shell that is either biconvex or plano-convex, and they bear a long straight hinge and a triangular hinge area on both of the valves. The brachial apparatus is confined to crura, or processes upon which the soft parts were suspended. They are found only in the Palaeozoic era, and lived from the Lower Cambrian to the Upper Permian period.

The genus *Orthis*, which dates from the Ordovician period, is typical of this order, and takes its name from it. The organisms have a shell with a long straight hinge, and are roughly square, or oval, planoconvex, and covered on both valves with large ridges or ribs that radiate from the umbo and give it the appearance of a *Pecten* (bivalve mollusc).

The order Strophomenida is the largest order of brachiopods, consisting of about 400 genera that lived from the Lower Ordovician to the Lower Jurassic period. They vary widely in shape from the typical ones that have roughly semicircular concavo-convex and flat-convex shells to those modified for attachment to a coral reef, and finally those that develop spiny processes used for attaching themselves to the sea bottom without use of the pedicle. Most species have a well-developed hinge and a pedicle aperture at the top of the umbo, but lack a spondylium. Their muscles are therefore inserted directly into the internal walls of the valves.

The genus *Gemmellaroia*, found in Permian reefs off Russia and Sicily, is a typical example of shell modification to suit a particular set of environmental conditions. It bears a remarkable similarity in shape to acephalous Cretaceous lamellibranchs that populated the same marine surroundings millions of years later. In the *Gemmellaroia* the ventral valve has become an elongated and irregular cone, while the dorsal valve is transformed into a virtually circular operculum. As in all reef animals, the shell wall is thick and robust and is formed by the three superimposed layers that are common to all the brachiopods. Typical of these is the genus *Productus*, which appeared in the Lower Carboniferous era.

The Pentamerida comprises brachiopods that lived from the Middle Cambrian to the Upper Devonian period, with a biconvex shell and a curved hinge edge, equipped with crura and a spondylium. The most common genus is *Pentamerus*, a guide fossil of the Silurian period, which has an almost circular biconvex shell that may have either smooth or ornamented valves with delicate sparse radial edges.

The order Rhynchonellida includes a large number of brachiopods that date back to the middle of the Ordovician period, some representatives of which are still to be found alive today. They have shells with arched valves, equipped with a broad protuberance on the ventral valve and a corresponding one on the opposite valve. The hinge edge is curved, and inside the lophophore is supported by well developed crura. At one time the 600 different forms were classified in the genus *Rhynchonella*; now this name is reserved solely for certain species of the Upper Jurassic and Lower Cretaceous periods. Strictly speaking, the genus *Rhynchonella* is characterized by a smooth triangular shell with a high dorsal fold and a deep ventral furrow, while other representatives of this large order tend to possess a spherical or triangular shell, ornamented by strong radial ridges which have a zig-zag edge at the front, while the deep ventral recess and the corresponding dorsal bulge are marked on the line of junction of the valves by a break in the line itself.

The Spiriferida are articulate brachiopods that existed from the Middle Ordovician to the Jurassic period. Their main feature is the presence inside the valves of a spiral brachidium, which in the living animal supported a lophophore of the same shape. The genus *Spirifer*, which has given its name to the entire order, has a very characteristic shell with a long straight hinge edge, marked radial ridges, a highly developed umbo on the ventral valve, and a broad protuberance on the valve itself. *Spirifer* lived during the Carboniferous period and may be found in large numbers in all the marine terrains of that epoch.

The order of the Terebratulida includes some very old forms, dating back to the Lower Devonian period, and some genera that are alive now. These last include *Gryphus* and *Argyrotheca*, which are widely distributed in the Mediterranean, and *Magellania*, only found in the Southern Hemisphere. All the representatives of this group are characterized by thin brachia which can at times be very complex. They have furnished a large number of guide fossils for palaeontologists all over the world.

Among the principal genera is *Terebratula*, characteristic of the Miocene and Pliocene periods. The shape of this species is said to have given rise to the name 'lamp-shells'.

# The Worms

The animals that we commonly call worms are so varied in their anatomy that their classification becomes particularly difficult. It is principally based upon the characteristics of the soft part of the body, and these have made it possible to subdivide the animals into phyla, each phylum differing significantly from the next. To speak of 'worms' in palaeontology would at first seem absurd, since these animals possess soft bodies, for the most part entirely unprovided with hard parts and thus difficult to preserve. Evidence of worms is generally found in the form of impressions, without anatomical details, or in the form of the tracks that they left in moving over the Earth, or even in the form of the habitations that they constructed.

Naturally these traces do not give us much clue as to the appearance of the animal, and for this reason palaeontologists tend to overlook the existence of the majority of worms – although they were widespread in the oldest geological eras – and limit themselves to recognizing only those that have constructed habitations and left traces, or that possessed anatomical details preserved for us in the most resistant of materials. However, long years of research carried out in sedimentary rocks have furnished a mass of fossil remains that can be attributed to various phyla of this animal group, and which represent only a part of the various groups living today, some of which are not to be found in Earth's rock layers, even

though they are thought to be very old.

Nevertheless, worms are fairly common in the fossil state, particularly in sedimentary marine rocks deposited at shallow depths. Some specimens are known from Precambrian times, mainly in the form of impressions or of tubes excavated in the sandy or muddy bottoms. Impressions of the body or habitation tubes only begin to appear much later, in the Cambrian period. Small fossilized mandibles are found for the first time in the Ordovician period, while worms are found in freshwater deposits only in the later periods. The known fossil traces can be attributed to at least six different phyla: some, such as nemertines (Nemerta), nematomorphs (Nematomorpha) and nematodes (Nematoda), are known in the fossil state only from the Jurassic or the Tertiary periods; others, such as the annelida (Annelida) are much less restricted; and yet others, the chaetognaths (Chaetognatha), are considered by some zoologists to be much closer to the echinoderms than to the annelids. Finally the sipunculoids (Sipunculoida), are both interesting and of great antiquity.

Of all the worms, however, those that are of most interest to the collector are the annelids, traces of which are to be found with great frequency in the rocks of all the eras, from the

*Above: This imprint, discovered on the surface of a stratum of calcareous marl from the Tertiary beds of the Apennines, was probably caused by a limivore worm filling up a burrow in the muddy sea bed (much reduced). Far left: Living tubes of Serpula sp., an annelid belonging to the Polychaetia, from the Lower Jurassic, Germany (lifesize)*

Cambrian to our own day (a period of about 500
million years). They are subdivided into a number
of different classes, but here we shall mention
only the class Polychaeta, which appeared in the
Cambrian period and is subdivided into the
errants (Errantida) which have few fossil representatives, the sedentaries (Sedentarida), which
are much more common, and Miskoiida, which
include some of the strangest invertebrates.

Leaving aside the errant polychaetes, an order
of soft-bodied worms whose traces are found in
terrains of the Ordovician period, we will turn
to the sedentary polychaetes. These have left
remarkable traces of their existence in the form
of tubes, for the most part calcareous, or of
holes excavated in the sea bottom, which they
used as habitations. They are known from the
Cambrian period, and the oldest representatives,
which existed about 500 million years ago, do not
seem to differ greatly from the many living types.

Among the best known, however, is the genus
*Serpula* which dates from the Silurian period.
The animal builds itself a twisted or straight
calcareous tube, which is attached at the rear to
any fixed object, such as a submerged rock or a
mollusc shell. The genus *Hicetes* also constructs
a small tube, which varies in diameter and shape,
and is fixed to the top of the coral *Pleurodictyum*

| Phylum | Class | Order | Age |
|---|---|---|---|
| Nemerta<br>Nematomorpha<br>Nematoida<br>Chaetognatha | | | Jurassic-Recent<br>Eocene-Recent<br>Oligocene-Recent<br>Cambrian-Recent |
| Anellida | Polychaeta | Errantida<br>Sedentarida<br>Miskoiida | Ordovician-Recent<br>Cambrian-Recent<br>Cambrian-Recent |
| | Myzostomia<br>Oligochaeta | | Ordovician-Recent<br>Carboniferous-Recent |
| Sipunculoida | | | Cambrian-Recent |

have been perfectly preserved in the clay.

The phylum Chaetognatha today includes no more than 30 marine genera, which are important in the constitution of plankton. They are in fact 'worms' adapted for swimming, and have developed caudal and lateral fins. The single fossil representative of this phylum, which is mentioned here only because it is rare, is the genus *Amiskwia*, found in the same Cambrian layer. It is very similar to the chaetognatha that are to be found today, and shows that the group had reached its present evolutionary stage as early as the Cambrian period.

The conodonts are fossils resembling the scolecodonts of the errant polychaetes. They are microfossils, a few millimetres long, in the shape of mandible plates, common in sedimentary rocks from the Ordovician to the Triassic period. The conodonts however differ from the chitino-siliceous scoleodonts, in that they are formed of calcium phosphate and, unlike these last, have a wide distribution, being important to zoologists for their usefulness in determining the relative ages of the rock strata.

Their classification as 'worms' is, however, still tentative, and indeed they were at one time attributed to the molluscs, because of the resemblance that they bear to mollusc radulae, and to the arthropods, and even to some unknown primitive vertebrates.

Notwithstanding such problems of classification, however, the conodonts are to be found in many Palaeozoic and Triassic sedimentary rocks. Their remains, to which difficult names are given, can be seen in microscopic sections of many Palaeozoic limestones in Europe and North America.

*problematicum*, with which it shared a symbiotic existence during the Devonian period.

Finally, a few genera found in the Cambrian layers of the Burgess clay in British Columbia are attributed to the order Miskoiida. They are marine organisms, mostly large in size, whose structure may be studied in detail because they

*Below: A trace of an unidentified marine worm from the Miocene rock of Däniken, Switzerland*

48

# The Arthropods

The phylum of the arthropods (Arthropoda) comprises an extremely broad and varied group of invertebrates, nowadays the largest phylum of the animal kingdom, the history of which began many aeons ago in the Precambrian period. The fossil representatives of the phylum vary widely in size, from microscopic insects, less than 0.25 centimetres in length, to the impressive giant eurypterids, which may be over two metres (six feet) long. They are more or less common in all sedimentary rocks, from the Cambrian period until today. Some, like the trilobites and the ostracods, are extremely common, and serve as important guide fossils for numerous geological periods. Others, like the malacostraca and the

and the functions that they perform. Thus appendages on the head may have developed into antennae, claws, or mandibles; thoracic appendages may be used for locomotion, and abdominal ones for respiration or swimming.

Leaving aside the internal anatomy, which is of little interest to palaeontologists, since only the skeleton is preserved, the most important single characteristic is that the fossilized arthropods were at one time exclusively aquatic animals, and it was only much later that some of them became terrestrial, and still others became adapted for flight. The oldest fossil classified as an arthropod was found in Precambrian terrains and suggests a long history for this phylum.

*Like all arthropods, the trilobites periodically shed their carapace. For this reason sedimentary rocks tend to contain only incomplete portions of the exoskeleton. This picture shows the pygidia of two separate trilobites, discarded after moulting. Upper Silurian; Kosor, Bohemia (×1.5)*

insects, need particular conditions to preserve their delicate structures, and are thus very rare indeed in the fossilized state.

The arthropods are animals that have an external chitinous or chitino-phosphatic skeleton, segmented and articulated, and typically formed by three principal parts: the head, the thorax, and the abdomen. Each one of these is in its turn formed by a certain number of segments, sometimes welded together, and each originally provided with two appendages. These appendages, while common to all the primitive arthropods, differ among the more evolved types according to the part of the body in which they are situated

The first arthropods that can be called such with any degree of certainty, come from Cambrian rocks, notably the Burgess shales in British Columbia, which have yielded some of the organisms discussed and are again of great importance in regard to these animals. On the whole, however, the evolutionary history of the arthropods is difficult to establish, because fossils that are sufficiently old are lacking. Basic functional and structural resemblances between two classes are often taken to be evidence of a relationship between them, in the absence of more detailed information. In point of fact, however, the Burgess strata have provided a

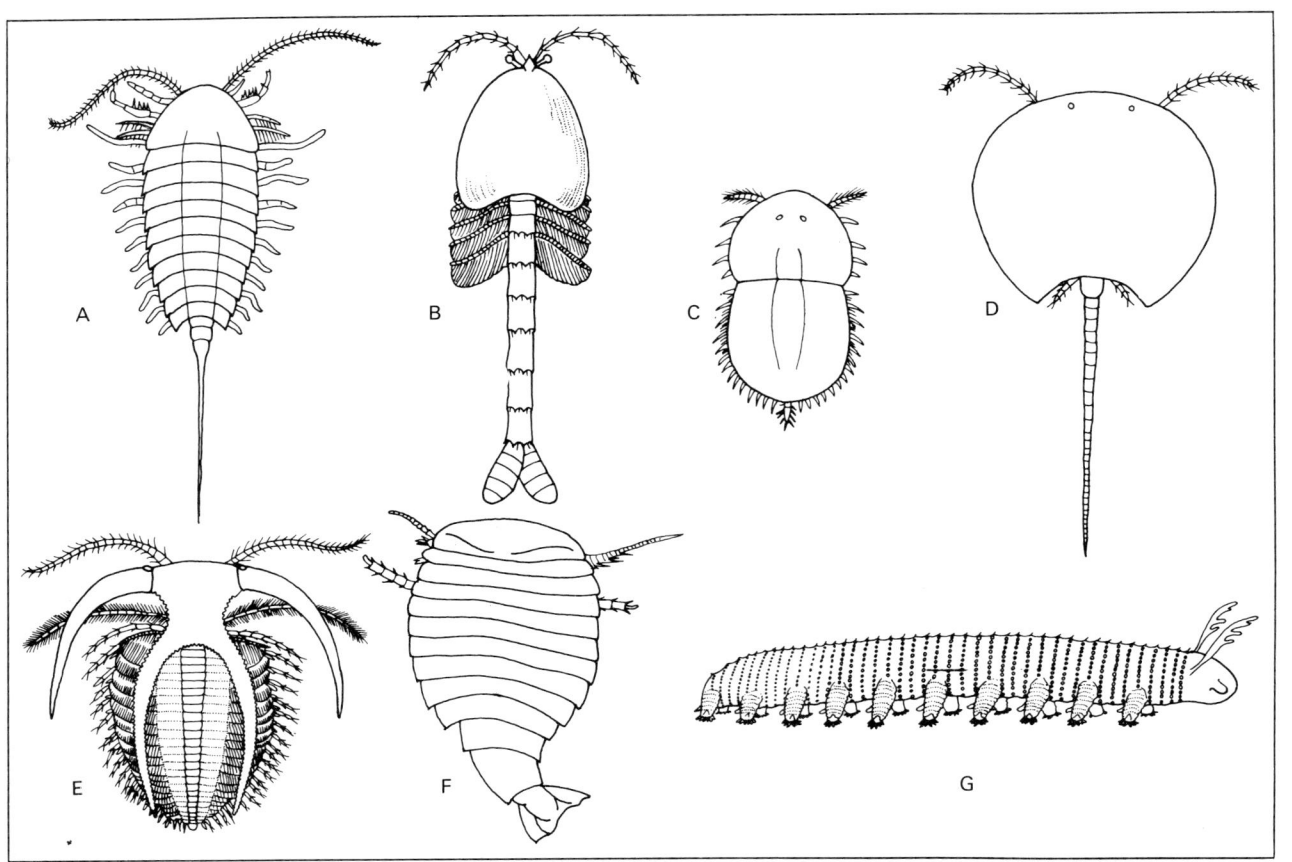

number of species believed to be transitional between one group of arthropods and another, and these species greatly assist determination of the links between the various groups as well as the descent of the arthropods from the annelids.

It is believed that the arthropods evolved from segmented aquatic worms similar to annelids during the Precambrian period. The arthropods have inherited from these organisms the structure of their nervous system, the method of growth of their segments and the characteristic elongated body shape, but have in addition developed an external skeleton, which has meant loss of the flexibility and the ability to contract that was possessed by their ancestors. Before examining these curious types in more detail however, it should be said that a recent palaeontological classification divides the arthropods into two main groups: the protoarthropods (Protoarthropoda), which include among others the fascinating onycophores (Onycophora), the only ones known in the fossil state; and the euarthropods (Euarthropoda), including trilobitomorphs (Trilobitomorpha), chelicerates (Chelicerata), and mandibulata.

## Subphylum Onycophora
The fertile Cambrian strata of the Burgess shales have yielded fossils of the genus *Aysheaia*, which are attributed to the onycophores, and believed to be their oldest representative, as well as one of the most primitive arthropods known. Palaeomtologists are not in agreement over the position occupied by the onycophores in arthropod classification, and they are either considered to be a separate phylum or attributed to the protoarthropods, a primitive and rare group of fossil representatives. The body has a large number of small legs and is slightly similar to an annelid in its marked segmentation and the undifferentiated head, but, unlike the annelid, it possesses a pair of short antannae, small eyes and a mouth with two lateral mandibles. *Aysheaia* was a marine animal, in contrast to the present-day onycophores, which live in tropical forests.

## Subphylum Trilobitomorpha
The Trilobitomorpha are the oldest of the euarthropods, comprising two classes of organisms, the Trilobitoidea and the Trilobita, which were extinct by the end of the Palaeozoic era and are thus known only through their fossilized remains.

### Trilobitoidea
The class Trilobitoidea groups together some Cambrian organisms that possess details which permit us to consider them the forerunners of a number of different groups of arthropods; groups that are to be found well developed in later periods of Earth's history. Among these are the Merostomoidea, to which belong the genera

| Phylum | Supersubphylum | Subphylum | Class | Age |
|---|---|---|---|---|
| | Protarthropoda | Onychophora | | Precambrian-Recent |
| Arthropoda | Euarthropoda | Trilobitomorpha | Trilobitoidea Trilobita | Cambrian-Devonian Cambrian-Permian |
| | | Chelicerata Pycngonida | Merostomata Arachnida | Cambrian-Recent Silurian-Recent Devonian-Recent |
| | | Mandibulata | Crustacea Myriapoda Insecta | Cambrian-Recent Silurian-Recent Devonian-Recent |

*Conocoryphe sulzeri, a small trilobite of the order Ptychopariida, from the Middle Cambrian of Bohemia (×5)*

*Sidneyia, Naraoia,* and *Emeraldella.* These are considered by some to be the ancestors of the merostomes, a group that appeared in the Upper Silurian period, because of the trilobate dorsal shield that partially covers the body, and its primitive appendages and slim elongated pygidium. The genus *Sidneyia* in fact, closely resembles the euripteride merostomes, while *Naraoia* is reminiscent of the xiphosures.

The Pseudonotostraca, other trilobitoids, still known as pseudo-crustaceans by some scientists, in many ways recall the crustaceans, and were indeed classified as such because they lacked the trilobation of the carapace that is characteristic of true trilobites. Others, notably the small genus *Burgessia* and the genus *Waptia,* are very similar

*Above left:* Elrathia
kingi, *a trilobite of
the Middle Cambrian,
Utah. Length 2.7 cm.
Upper right:*
Triplagnostus
burgessensis, *a
primitive trilobite
of the order Agnostida,
with only two abdominal
segments and with
cephalon and pygidium
almost equal. From the
Burgess clays,
British Columbia.
Length 0.6 cm. Lower
right: Broken off
pygidium of*
Odontochile hausmanni
*Lower Devonian;
Lochkov, Bohemia.
Slightly enlarged*

*Left: Other trilobites
of the Burgess clays:
1.* Ogygopis klotzi;
*2.* Olenoides serratus

52

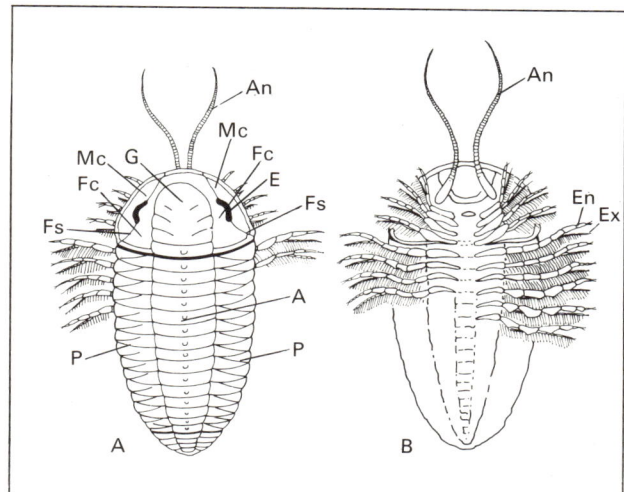

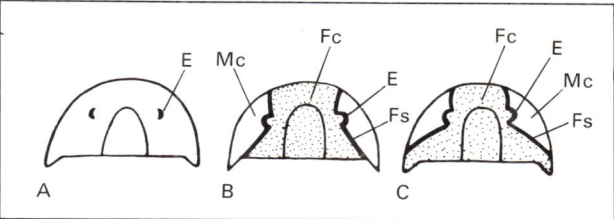

division of the body into three parts. The longitudinal sections are a central part, or rachis, and two lateral parts, known as pleurae. The transverse lobes are formed by a cephalic shield (cephalon) on which are situated the kidney-shaped compound eyes, a central swelling (glabella) and a facial suture. The position of this

to the branchiopods, and would be classified as such, but for the fact that they possess primitive appendages. Finally the Marrellomorpha, to which the genus *Marella* belongs, are so strange in appearance that they are not attributed to any group.

suture, which seems to have divided the fixed central part of the shield from movable lateral parts, is a crucial element in the classification of these organisms.

The abdomen lies behind the cephalon, and is composed of segments that articulate with each other, varying in number between two and twenty-two. The third lobe, situated at the rear end of the body, is a triangular part known as the pygidium, which is formed by the fusion of a number of segments, and sometimes bears a long spine at its end. The undersurface of the animal occasionally displays a number of appendages,

## Trilobita

The trilobites, fossils that are sought by collectors all over the world, belong to a class of primitive marine arthropods that lived exclusively during the Palaeozoic era. The name comes from the longitudinal and transverse

which are primitive in comparison with those of other arthropods. Each segment of the trilobite carries two appendages which – apart from the first pair, which have become antennae – are identical. They are undifferentiated, and each consist of a basic part (protopodite) which carries an endopodite and a ciliate exopodite, the first probably used for swimming and the second for crawling. The cephalon, formed from the fusion of five segments which are no longer distinguishable, carries five pairs of appendages.

The trilobites are first known from the Cambrian period, in which they were already greatly evolved. But this degree of evolution indicates a much older history, with its roots buried far back in the Archaeozoic era. The trilobites multiplied in the following period, and in the Silurian they became more and more common and developed the ability to roll themselves up in defence.

The descendant phase begins in the Devonian period, and by the Carboniferous period only two families are to be found, and these in turn disappeared at the end of the Permian period.

The trilobites are found in marine deposits in very varied surroundings, a fact that indicates how versatile these animals were in adapting to different environmental conditions. The flattened form of the body seems, however, to indicate

*Above: A large specimen of* Olenellus thompsoni, *from the Lower Cambrian of Mt White, British Columbia (slightly reduced). Left: Another specimen of* Olenoides serratus, *one of the most characteristic trilobites of the Cambrian. Mt Stephen, Canada. Length of the original, 3.7 cm.*

*Right: A perfect specimen, much enlarged, of* Phacops rana, *from the Devonian levels of Ohio*

benthonic habitats for the majority of the types, and it is thought that they lived on the sea bottom, and dug themselves into it in the same way as the present-day *Limulus*, the king-crabs. Other types seem instead to be adapted to the nectonian life or an epiplanktonic existence, particularly the species provided with huge eyes and an inflated glabella which appears to have served to keep the organism afloat. The strata of Scandinavia, England, Russia, Bohemia and North America, countries in which there are considerable Palaeozoic outcrops, are very rich in trilobites.

The class trilobites is subdivided into various orders which comprise a large number of very varied groups. To cite only the principal ones: there are the Agnostida, probably the most primitive, which were alive from the Lower

Cambrian to the Ordovician eras. They lack a facial suture, but have a highly developed cephalon and pygidium, which are almost identical, and have only two or three abdominal segments. The Redlichiida from the Lower and Middle Cambrian period are characterized by well-developed lateral points (genal points) and by an opisthoparian facial suture; this passes behind the genal points, which thus form part of the movable section of the cephalon. The Corynexochida tend to be elliptical in shape, with a semicircular cephalon, and bear an opisthoparian suture or seam. Characteristic of the Postcambrian trilobites are the Phacopida, which have a 'proparian' suture in which the genal spines are to be found in the fixed part of the cephalon. Finally, the Ptychopariida, trilobites

*During the Silurian era trilobites acquired the ability to curl themselves up, probably for defensive purposes; the illustration shows a much enlarged specimen of Synhomalonotus tristani (order Phacopida) from the Lower Silurian, Ciudad Real, Spain*

with an opisthoparian suture, have an oval exoskeleton and are elongated, with between 12 and 17 abdominal segments. They were very widely distributed from the Lower Cambrian to the Ordovician Period, and continued living through the Palaeozoic era until the Permian period, when the only two families surviving, Proetidae and Phillipsidae, became extinct.

## Subphylum Chelicerata

The chelicerata include both terrestrial and aquatic arthropods known from the end of the Cambrian period and grouped into the following classes: the arachnids (Arachnida), merostomes (Merostomata) and pycnogonids (Pycnogonida). The arachnids, which are rarely found as fossils, are principally terrestrial, only a few species being adapted to an aquatic environment; the merostomes, which are much more common, are exclusively aquatic animals, while the pycnogonids, which are extremely rare in the fossil state, are only to be found in the sea.

The interest of the arachnids to palaeontology is primarily due to the fact that they include the first animal that left the waters to live on land. This is the 'scorpion', *Palaeophonus nuncius*, which moved on to the deserted continents during the Upper Silurian period. The 'spiders', to be found for the first time in Carboniferous strata, are however much younger, and lived in large numbers in the immense forests which are known to us as vast deposits of coal. As we shall see with

*Right:* Olenellus thompsoni, *a trilobite of the order Redlichiida, from the Lower Cambrian, British Columbia (slightly enlarged). On the stone only the remains of the cephalon are visible*

*Below:* Lloydolithus ornatus, *a trilobite of the order Ptychopariida, from the Middle Ordovician. It is characterized by the greatly developed and low-lying cephalic margin, and by two long spines. Britain*

the insects, numerous specimens of arachnids are preserved in amber, the fossilized resin of an ancient Tertiary conifer.

The representatives of the merostomes are, however, much more strange and unusual. This class is divided further into two subclasses: the eurypterides (Eurypterida) and the Xiphosuria. The Euripterida, which are also known as *gigantostraci* because of the enormous size of some species, lived from the Silurian to the Permian period, at first in marine waters, then in estuarine and inland waters. They had an elongated body formed by a cephalic shield with two lateral and two secondary eyes, and, underneath, six pairs of appendices, the first equipped with claws and the others developed to enable the organism to walk or swim. The abdomen was made up of segments, the telson elongated.

The best-known genera are *Pterygotus*, the largest arthropod that existed (some 2½ feet long) from the Upper Silurian and Devonian periods, and *Eurypterus*, which was very much smaller, and was to be found up to the Carboniferous period.

The Xiphosuria are even more familar through

the limuli (king-crabs), which appeared during the Triassic period, 170 million years ago, and are still to be found in the coastal waters of North America, Japan, China and Indo-China.

## Subphylum Mandibulata

The Mandibulata include the most advanced arthropod species. They appeared on the Earth during the Cambrian period, and over the ensuing 500 million years evolved into a group containing species adapted to every sort of environment. They are divided into three large classes: crustaceans (Crustacea), myriapods (Myriapoda) and insects (Insecta).

### Crustacea

Together with the insects, the crustaceans are the most important arthropods living, to be found in great numbers in seas and continental fresh waters; the presence of a strong mineralized carapace has meant that they are relatively common in the fossil state. The history of their evolution, which started in the Cambrian period, has continued until today with a few changes.

They are grouped in five subclasses: branchiopods (Branchiopoda), ostracods (Ostracoda), copepods (Copepoda), cirripeds (Cirripedia) and malacostraca (Malacostraca). Of these, only the ostracods, the cirripedes and the malacostraca are of interest to the amateur fossil collector.

The ostracods are small crustacea that fossilize well, due to their calcareous or chitinous bivalve shell. Because they are common in many types of marine rocks, they are important as guide fossils for the dating of Palaeozoic and Mesozoic terrains. The method of collecting and studying these minute animals is similar to that employed for the foraminifera and the conodonts.

The cirripeds (barnacles) are strange crustacea whose bodies are protected by a thick calcareous 'shell' formed by overlapping plates. They are of little chronological value, but are useful to palaeontologists as ecological indicators. They are only too familiar today living in coastal waters, attached to rocks or other organisms and their presence indicates the vicinity of a coast line.

Finally, the malacostraca. Of these we shall

*Above: Two specimens of* Dalmanitina socialis, *a trilobite of the order Phacopida, from the Ordovician, Bohemia. Length of the specimens, 6.7 cm*

*Right: Ellipsocephalus hoffi, a Cambrian trilobite of the order Redlichiida, from Czechoslovakia. Length of the original, 3.3 cm*

mention here only the order of the Decapoda, which includes the most advanced species including lobsters and crabs, which are easily fossilized due to their hard carapace. They are divided into Macrura, which have an elongated abdomen and tail fin, Paguridae, which have a soft abdomen that is fossilized only with difficulty, and Brachyura, which has a small abdomen folded under the body. Of the macrura, which appeared during the Triassic period 170 million years ago, the Eryonides are particularly notable. Excellent examples of these animals come from the Upper Jurassic layers of Solnhofen in Germany, a mine of perfectly preserved fossil forms, while even older eryonides, dating back to the Lower Lias period, have recently been found on the Italian side of Lake Lugano and have been attributed to the genus *Coleia*.

Of the many forms which are represented by this order, numerous species of lobster and crab are to be found from Lower Palaeozoic to recent times. Many of these species can be obtained as fossils from the Mesozoic and Tertiary soft-rocks of California, Texas, Panama and numerous European localities.

## Myriapoda

Little is known of the myriapod fossils, for their delicate structure is not conducive to good fossilization. Zoologists divide the group into four classes, all of which appeared in the Silurian period, developed during the Carboniferous period and continued evolving until the present day, changing little over the millions of years. The best fossil examples come from Oligocene layers, in which these delicate animals are imprisoned and preserved for all time in amber. Millipedes and centipedes are still very common.

*Above left:*
Eurypterid remipes, *a eurypterid Merostomata from the Upper Silurian, New York. Length of the original, 8.1 cm.*
*Above right:* Eurypterus lacustris, *another eurypterid of the Upper Silurian, New York (lifesize)*

*Right: A crab, 60 million years old, perfectly preserved in the Eocene limestone of Monte Bolca, Verona*

Left: Specimens of Balanus *sp., on a valve of* Flabellipecten. *From the Pliocene, Altavilla, Sicily. Size of the shell, 11 cm*

Below left: Drawing of Palaeophonus nuncius, *a 'scorpion' of the Silurian, and the first animal to live on the land.* Below right: Drawing of Pterygotus, *a eurypterid of the Merostomata with a length of more than 2 metres, which lived in continental fresh water during the Silurian era.* Right: Mesolimulus walchii, *a merostome of the Xiphosura, and the direct ancestor of the present-day* Limulus. *From the Upper Jurassic; Solnhofen, Germany (slightly reduced)*

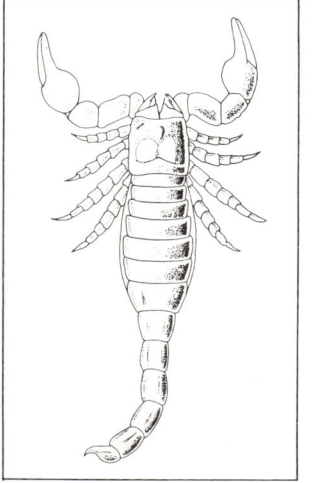

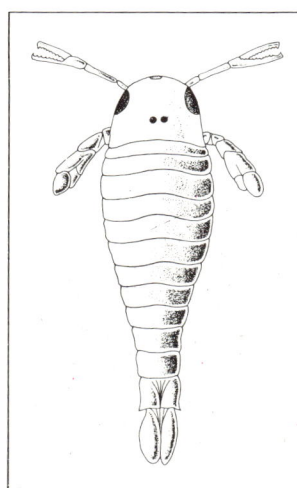

## Insecta

The insects, or Hexapoda, are arthropods that are seldom found as fossils, since they fossilize only in sediment that is fine enough and consolidates quickly enough to provide adequate protection from the atmosphere. Species preserved in amber are thus very important, and constitute a precious source of perfectly preserved specimens. As happens in pine forests of today, numerous insects as well as other invertebrates such as myriapods and spiders, become encapsulated in falling resin, which has since fossilized. This amber generally dates back to the Oligocene period, and was known to the ancients, who, unable to account for the animals enclosed inside it, attributed magic properties to the substance. It has furnished us large numbers of perfect examples of the ants, termites, bees, beetles and other insects alive at that time.

*Far left: Another brachiura, probably from the Oligocene. Vincentino (×2). Below left: Palinurus sp., a lobster of 60 million years ago. From the Eocene of Monte Bolca, Verona. Length, 33 cm. Below right: Coleia viallii, a very rare erionid from the Lower Jurassic; Osteno, Como. Erionids are crustaceans which have now disappeared (slightly enlarged)*

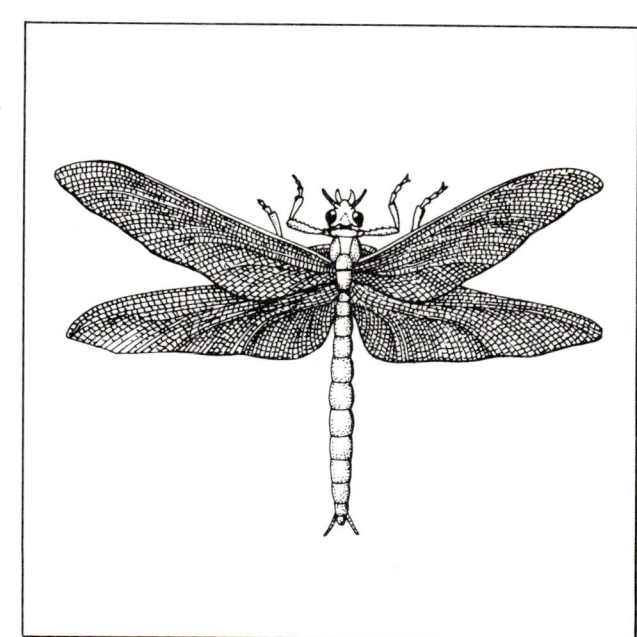

*Left:* Coleia sp., *another decapod crustacean, from the new and unusual Osteno bed on Lake Lugano. Lower Jurassic. Length of the original, 3.3 cm*

*Right: Reconstruction of Meganeura, a gigantic protodithyopter of the Carboniferous era. Below:* Eryon arctiformis, *an eryonid crustacean of the Upper Jurassic. Solnhofen, Germany. Length of the specimen, 11 cm*

Insect remains also exist in certain marine deposits, having been carried there by the wind. They were fossilized along with other aquatic animals. This is the case, for example, in the Eocene of Monte Bolca near Verona, where insect remains are found perfectly preserved beside crustaceans and fish of the same era, some of them still showing the original colours, preserved as in the living insect.

The fossil insects of Solnhofen in Bavaria are even better preserved; this Jurassic limestone has permitted the preservation of the most delicate structures because of the very fine grain of the deposit. A fine example of this preservation are the beautiful dragon-flies, whose fine membranous wings have left wonderfully precise impressions in the rock. Similar, if less complete, insects have been found in the still earlier rocks of Dorset, in Britain.

The evolution of the insects is very controversial; some scientists believe that they derive from the trilobites, others that they came from the crustaceans, and yet others, from the annelids or from the myriapods. Nevertheless, what is certain is that the first insects to appear on the Earth were the Apterygota, which lack wings. Only later, in the Carboniferous period, did the first winged forms, or Pterygota appear. These were the Palaeodithyoptera, to which the genus *Maganeura*, the giant 'dragonfly' that has a wing span of 70 centimetres (28 inches) belongs. Still in the Carboniferous period, the Protoorthoptera, large ancestors of the cockroaches, the first grasshoppers, and the Hemiptera, all made their appearance.

The true dragon-flies or Odonata are first found rather later, in the Permian period, together with the Coleoptera (beetles), certain families of which

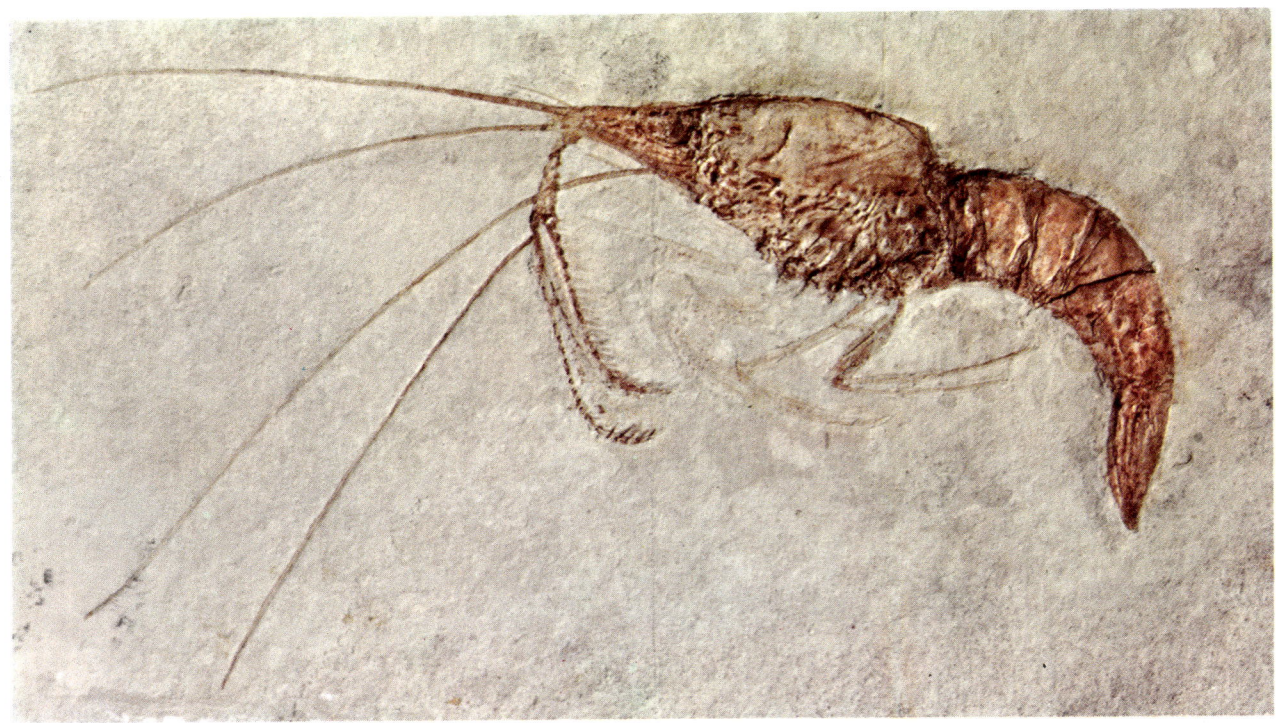

Aeger insignis, *a decapod crustacean of the Upper Jurassic; Solnhofen, Germany. Length of the original, 18.5 cm*

became very common in the Triassic and Jurassic periods, but are now nearly all extinct. The 'modern' coleoptera, as we know them today, developed in the Tertiary period. The Lepidoptera (butterflies and moths) and the Hymenoptera (bees) appeared as a result of the development of the flowering plants that occurred in the Jurassic period; notable among the latter group are the ants, which arrived during the Eocene period. The Diptera (flies) appeared at the beginning of the Jurassic period, and finally, to conclude our list, termites appeared during the Eocene period.

*Far left:* Phalnagites priscus, *a strange organism of the Upper Jurassic, holding the larva of a decapod crustacean. Eichstätt, Germany (×2.5).* *Left: A larva of* Libellula doris, *from the Miocene of Vittoria d'Alba, Cuneo, Italy (×2)*

# The Molluscs

With the molluscs we enter a vast and important group of invertebrates, whose members include animals as diverse as squids, octopuses, snails, slugs, mussels, clams, oysters and chitons. The name 'mollusc' is derived from the Latin word for 'soft', and describes the animal's soft body, which is often enclosed in a hard shell. Molluscs only have traces of the segmentation typical of the arthropods and annelids, and are generally defined as bilaterally symmetrical metazoa, although this symmetry is frequently hidden by torsion of the shell.

The typical mollusc has either an internal or an external shell, covered by a sheath that surrounds the viscera and which covers the shell itself. The soft body, which naturally is not found in the fossil state, is divided into three parts: the head, the feet and the viscera. Because the soft body is not fossilized, the molluscs are known to palaeon-

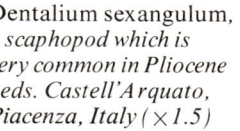

Dentalium sexangulum, a scaphopod which is very common in Pliocene beds. Castell'Arquato, Piacenza, Italy (×1.5)

tologists exclusively through the shell, which in the living animal is composed of aragonite, a mineral that transforms slowly into calcite during the fossilization. Classification is therefore based on the appearance of the shell, with each genus and species characterized by morphological details, such as the shape, the ornamentation, and the proportions of the shell. Classification into much larger groups, such as orders and families, is based on the features that are in some way linked to the soft structures – for instance, the imprints of the muscles, the trace of the passage of the siphon, or other parts. There exists, in fact, a certain divergence between the zoological classification of these animals, which is based almost exclusively on the internal anatomy of the animal, and the palaeontological classification, which is concerned with the little that fossilization has preserved of the animal. Such differences are lessening, however, as both sides learn more about the organisms, and palaeontologists and zoologists now tend to work together, seeking useful clues from the living species, as well as using the fossil forms as a source of knowledge about the origin of this animal phylum. The classification given in the table on page 66 is one of the several accepted classifications.

Naturally, the study of the most recent forms which are still in existence or are very close to the present forms, is relatively easy, because one may make comparisons and establish links. The study of the older groups that are now extinct is, on the other hand, considerably more complex, because they do not possess organs or an appearance that can be correlated with any present-day organism. This goes not only for the anatomy and appearance of the organisms, but also for what we know of the life-cycle and habitat of the molluscs, which are vitally important in the study of the rocks of the Earth's crust. The molluscs give useful hints about the conditions in which the sedimentary rocks were formed, once one knows their habits, the salinity, depths and temperatures in which they lived, and so on. For this reason they are excellent fossils for palaeographical reconstruction – that is, the reconstruction of the appearance of the land masses and the seas during the various geological epochs.

The evolutionary history of the molluscs is long and complex, because of the enormous variety of species that have succeeded one another in the course of the geological eras. They first made their appearance in the Lower Cambrian

| Phylum | Class | Subclass | Order | Age |
|---|---|---|---|---|
| Mollusca | Monoplacophora Aplacophora Polyplacophora Scaphopoda | | | Cambrian-Recent Recent Cambrian-Recent Ordovician-Recent |
| | Bivalvia | | Taxodontida Anisomyriida Eulamellibranchia | Cambrian-Recent Ordovician-Recent Silurian-Recent |
| | Gastropoda | Prosobranchia | Archaeogastropoda Mesogastropoda Neogastropoda | Cambrian-Recent Ordovician-Recent Ordovician-Recent |
| | | Opisthobrachia | | Devonian-Recent |
| | | Pulmonata | Basommatophora Stylommatophora | Carboniferous-Recent |
| | Cephalopoda | Nautiloidea Ammonoidea | | Cambrian-Recent Devonian-Cretaceous |
| | | Coleoidea | Belemnitida Sepiida Teuthida Octopodida | Carboniferous-Recent Jurassic-Recent Jurassic-Recent Cretaceous-Recent |

*Right: This association of beach molluscs, recently cemented together, is the first step in fossilization and the formation of an organic rock. Venice, the lagoon (× 1.5)*

period, some 500 million years ago, but their origin is thought to be still more remote. Mollusc evolution has in fact been singularly successful, and virtually all the classes are larger and more common than at any earlier stage in their history.

The molluscs are grouped into a phylum which is subdivided into six classes, comprising over 100,000 living species. Each class has a set of characteristics that is easily distinguished, even by amateurs. In recent years, a seventh class has been added, that of the monoplacophores. These last, at one time known only as fossils, were considered to be gastropods, but the discovery of specimens still living at the bottom of the Pacific Ocean has stimulated a more detailed study of the soft parts and the institution of these primitive organisms as a new class of molluscs.

The six other classes, and their main characteristics, are:

*Aplacophora*: no fossils;
*Polyplacophora*: body protected by a calcareous shell made up of articulated plates;
*Scaphopoda*: tubular shell;
*Bivalvia*: body protected by a bivalve shell;
*Gastropoda*: body protected by a spiral univalve shell;
*Cephalopoda*: sectate internal or external shell.

## Monoplacophora

As we have already said, the monoplacophores are a fascinating 'new' class whose members are considered the most primitive of all the molluscs. Until recently these animals were known exclu-

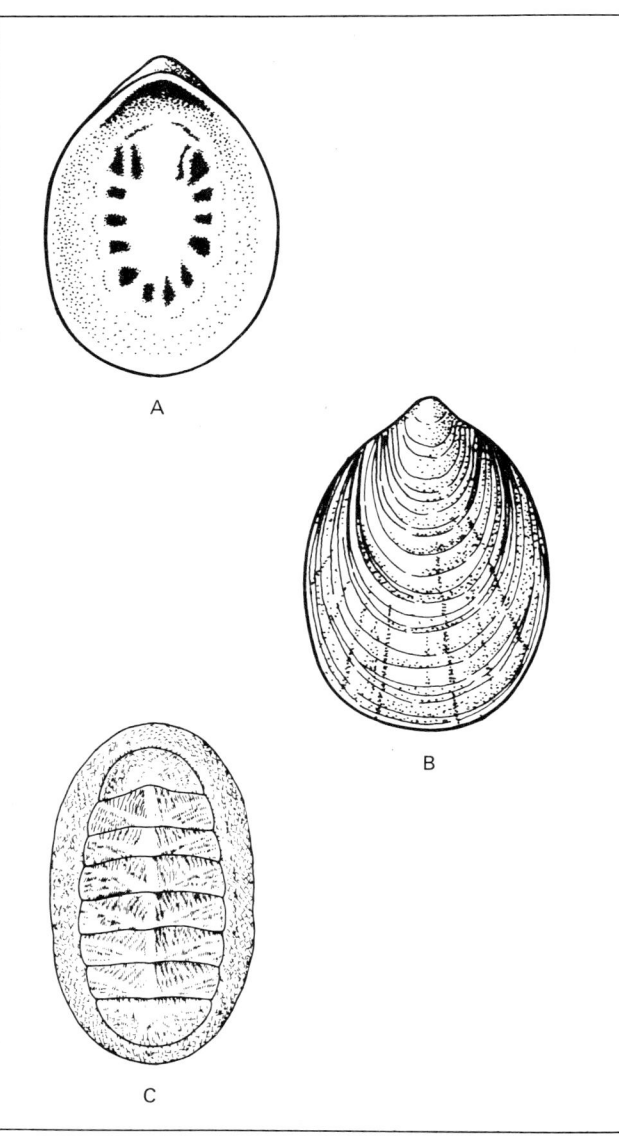

A

B

C

*Pilina sp., a monoplacophore mollusc of the Middle Silurian. A. Interior of the shell showing muscle scars; B. Exterior of one of the valves; C. Species of* Chiton, *a polyplacophore mollusc*

sively as fossils in rocks dating from the Lower Cambrian to the Middle Devonian period, and were considered to be completely extinct, until 1952, when a living monoplacophore, *Neopilina galatheae*, was dredged from the bottom of the Pacific Ocean at a depth of over 11,000 feet.

Typically, the monoplacophores have a univalve shell without spiralling, shaped like a cowl which protects the untwisted body. It is interesting to note that these animals have changed their habitat in the course of millions of years by moving from the continental shelves on which they lived 500 to 300 million years ago to the depths from which the *Neopilina* was dredged.

## Polyplacophora

The polyplacophores are marine molluscs with a bilaterally symmetrical body, equipped with a ventral 'foot', which is used for locomotion and often protected by a calcareous shell, or lorica (from which the alternative name of loricates is derived) formed by eight plates articulated one with another.

These are found as fossils from the Upper Cambrian period, mostly in the form of isolated pieces of the lorica; the genus *Chiton*, which appeared in the Cretaceous period and is still to be found today, is the best-known representative.

*Above: Association of* Claraia clarai, *a lamellibranch (or pelecypod) of the order Anisomyaria, quite common in the Triassic rocks of the Alps (life size). Left: Some oyster shells of the Anisomyaria, genus* Gryphaea. *From the Jurassic of Avallon, France (life size)*

*Right, top: Specimens of* Pteria contorta, *another of the Anisomyaria, in a piece of Triassic rock from Bavaria (×2.5). Below: Specimen of* Posidonia becheri *(Anisomyaria) in which both valves of the shell have been preserved. Lower Carboniferous; Lautenthal, Germany (×2)*

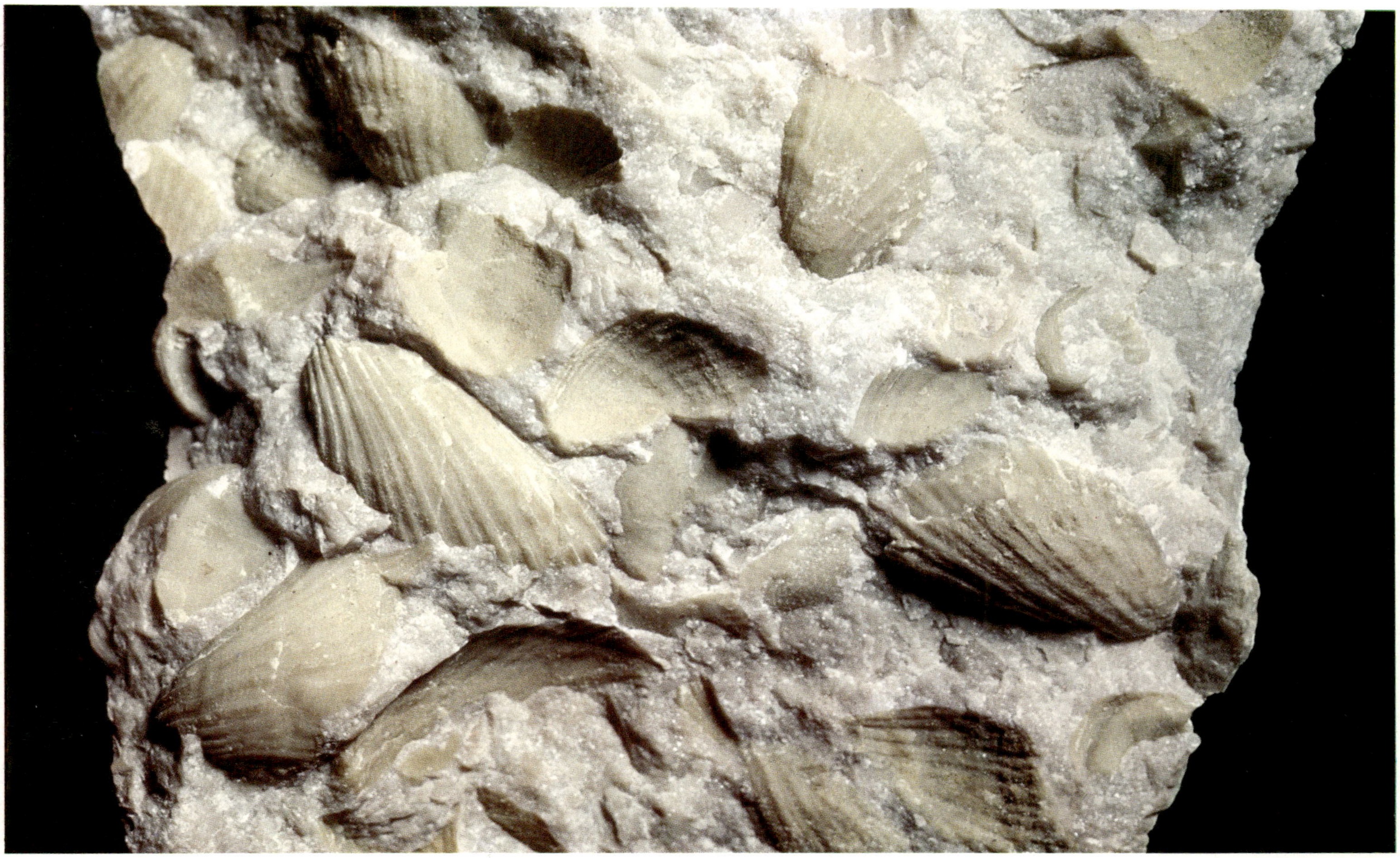

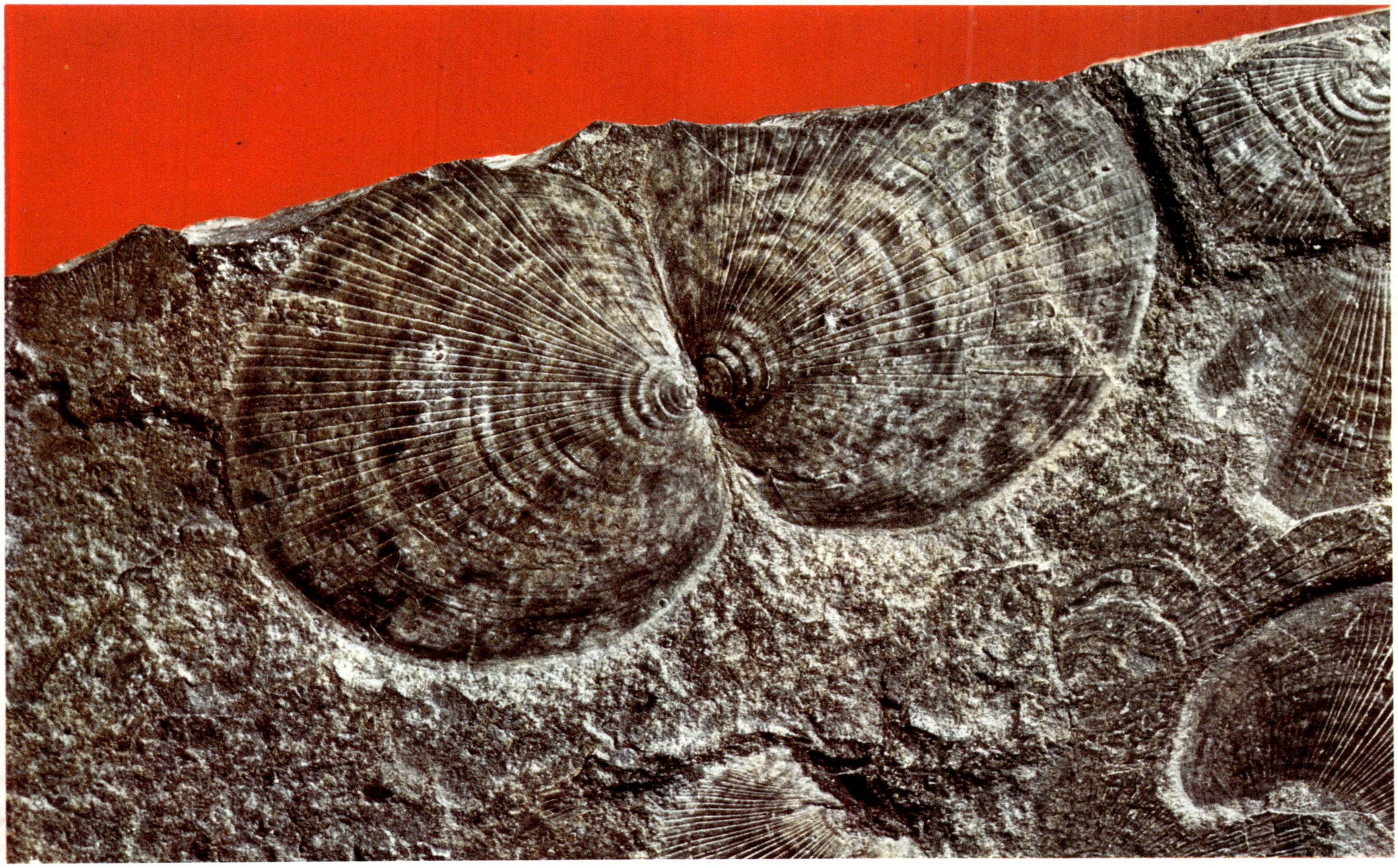

## Scaphopoda

The scaphopods are marine molluscs which have a univalve external shell in the form of a slightly curved tube that is open at both ends. For the purposes of description, the concave side of the shell is considered to be the dorsal face, the larger aperture as the anterior aperture, the smaller the posterior. The animal lives with its anterior part buried in the muddy bottom; only the posterior part, on which are situated both the genital apparatus and the animal's excretory mechanism, is in contact with the water.

The scaphopods appear to consist of very few species, largely because they all appear so much alike, but they are in fact very numerous, both today (no less than 200 living species are known), and as fossils. Indeed, a rough estimate of fossil species from the Ordovician period, when they first appeared, gives a tentative figure of 300 species.

## Bivalvia

The lamellibranchs, or bivalves, close to the affections of those who eat oysters, mussels and

*A Pecten of the Anisomyaria, very common in the Pliocene of Italy, Flabellipecten flabelliformis. Upper Pliocene; Val Andona, Asti, Italy (×1.5)*

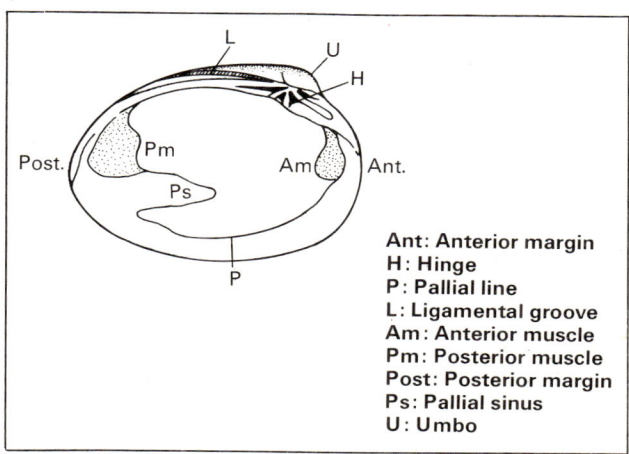

*Another Pecten of the Pliocene,* Flabellipecten cristatus, *from Balerna, Ticino, Switzerland (×1.5)*

| | |
|---|---|
| **Ant:** | **Anterior margin** |
| **H:** | **Hinge** |
| **P:** | **Pallial line** |
| **L:** | **Ligamental groove** |
| **Am:** | **Anterior muscle** |
| **Pm:** | **Posterior muscle** |
| **Post:** | **Posterior margin** |
| **Ps:** | **Pallial sinus** |
| **U:** | **Umbo** |

*The structure of a lamellibranch*

*Below left: Some of the variety of hinges in lamellibranchs: A. Taxodont (Glycymeris); B. Disodont (Pecten); C. Schizodont (Trigonia); D. Heterodont (Venus); E. Isodont (Spondylus); F. Desmodont (Mya); G. Pachiodont (Hippurites). (G1. Interior of the fixed valve; G2. Opercular valve). Below: Pedalion maxillatus, from the Pliocene of Val Andona, Asti, Italy. Height, 16 cm*

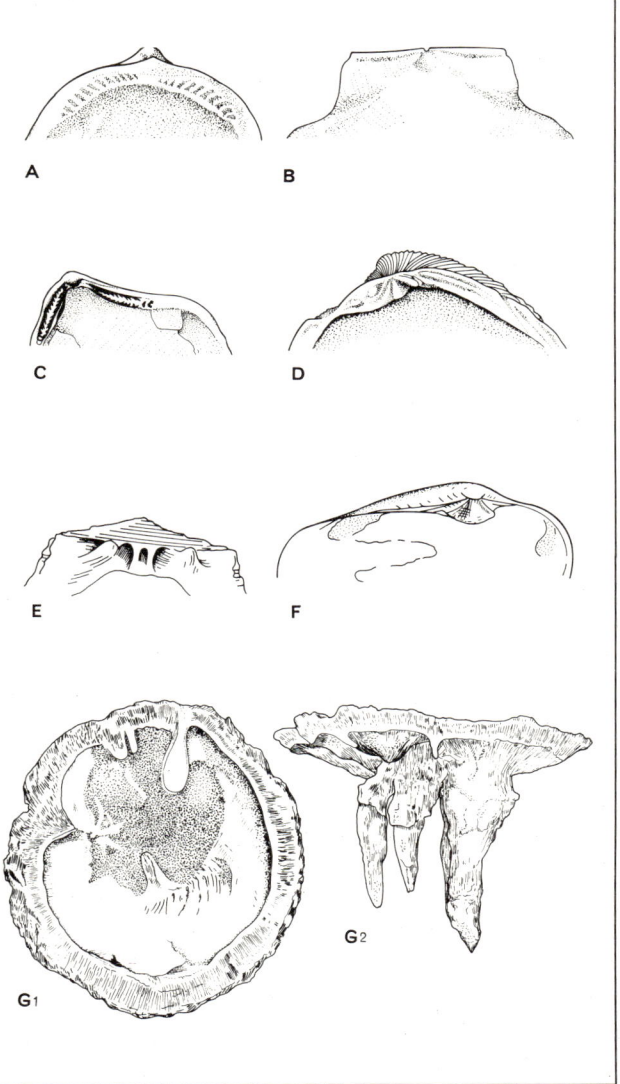

the many other similar types of seafood, are also of fundamental importance to the palaeontologist because of their wide distribution in all types of marine sediments, estuaries and freshwater deposits, and for the useful clues that they provide for him in his reconstruction of the Earth.

Their former name, Lamellibranchiata, derived from the particular shape of the branchia, the

*Above left:* Spondylus gaederopus, *an Anisomyaria which is widely distributed in Quaternary marine sediments. Archi, Reggio Calabria, Italy (slightly enlarged). Above right:* Megalodon gümbeli *in a rock section. Upper Triassic, Tremezzo, Como, Italy. Size of the original, 5.7 cm*

organ of respiration, which has a lamellar appearance. They are now however, called either bivalves, because their robust shell is formed by two valves, or pelecypods, from their flattened ventral foot, which is similar to an axe.

The external shell is formed, as we have said, by two valves hinged together, which enclose and protect the soft parts of the body of the animal. For palaeontologists the most interesting part, since it is the only one that is preserved in the fossil state, is the shell, and we shall describe that in detail here. Strictly the soft parts come into the domain of the zoologist, and need not concern us.

Equivalvular and equilateral in young specimens, the shell becomes non-equilateral in adult forms. The two valves frequently undergo variations or adaptations that modify their growth from the general pattern; thus in the types that have restricted mobility, such as *Pecten*, the shell is no longer equivalve, because one valve assumes an opercular shape. Still greater modifications are to be seen in the forms that attach to rocks – for example, *Ostrea*, in which the two valves are so unequal as to give a straight, elongated or conical shape.

The first step in the study of a lamellibranch is an estimation of the features, shape and size of the shell which then enables the scientist to des-

*Another specimen of* Megalodon gümbeli, *a lamellibranch used as an index fossil for the alpine beds of the Upper Triassic, Arcisate, Varese. Height of the original, 3.5 cm*

*Above:* Trigonia clavellata, *a lamellibranch of the Upper Jurassic. Weymouth, Britain. Size of the valve, 7.3 cm. Right: A complete specimen of* Myophoria kefersteini, *a typical lamellibranch of the Upper Triassic of the Alps. Lombardy foothills ( ×2 approx)*

cribe the anterior and posterior parts – an anterior muscle, a posterior muscle, and so on. Finally, the umbo, or curved part of the valves, is the feature that must be taken into account for an accurate description and classification of a lamellibranch. The umbo is found on the upper part of the shell, and in virtually all cases is bent towards the front, which is shorter than the posterior. Once these features are established it is easy to distinguish a front edge and a rear edge,

*Above:* Inoceramus balticus, *used as an index fossil for the beds of the Upper Cretaceous. Germany* (×1.5)

*Above: Some specimens of* Arctica islandica, *a typical lamellibranch of the cold climate of the Quaternary. Palermo (much reduced). Right: This lamellibranch, more than 70 million years old, still preserves some traces of its original shell. Upper Cretaceous; Germany (×2 approx)*

*Far left: Internal cast of* Pholadomya murchisoni. *Middle Jurassic; France (about lifesize). Left:* Isocardia cor, *a common lamellibranch of the Pliocene. Castell' Arquato, Piacenza, Italy. Size, 7.7 cm*

and to tell the right valve from the left valve.

Externally, the valves of the lamellibranchs, which are frequently coloured when the animal is alive, may be very varied in appearance, due to the ornamentation that covers them, which is an important feature in fossil classification. There are two principal types of decoration: either ridges and stripes that run round the umbo (concentric sculpture), or ridges and stripes that radiate from the umbo (radial sculpture). Additional nodes, spines and varices are often found as well. The growth of the shell takes place in alternating phases of rapid growth and a static state, and this causes concentric rings, known as growth rings, to appear on the outside of the shell. These are totally different from the ornamentation, and should not be confused with it.

Below the umbo, on the upper edge of the shell, is the ligament that holds the valves together, while below this is the hinge, which is formed on each valve by structures in relief (teeth) that engage in cavities (recesses) on the opposite valve, and which with the adductor muscles and the connecting ligament control the closing and opening movement of the valves, and prevent lateral movements.

Within the valves the soft parts leave signs on

*Left: Two specimens of* Hippurites radians, *a rudist lamellibranch of the Upper Cretaceous. Charente, France (lifesize).*
*Below:* Requienia ammonia, *a rudist of the Lower Cretaceous. France (lifesize)*

the shell which are used by palaeontologists in classification. These include the oval impressions left by the adductor muscles, which are situated at each side of both valves. (Sometimes the anterior muscle is lacking, and the lamellibranch is then said to be monomyarian.) These two muscular impressions are further joined by an indented line or pallial impression, which runs along the lower edge of the shell and is caused by the insertion of the muscles of the sheath. It is continuous, or, when the animal is equipped with a siphon, forms a large protuberance called the bladder locator, in the lower part of the valve. In the first case the mollusc is of the integral bladder type, in the second it has a located bladder.

There are, then, very few characteristics that have to be considered in classification of the bivalvia; the principal ones in connection with the internal structure are those that remain invariable, namely, the muscular imprint, the bladders and the hinge.

Numerous different types of hinge exist, of which the most important are:

*taxodont hinge:* a broad cardinal area that carries a large number of small regular teeth, separated by a series of recesses, the teeth and the recesses corresponding on both valves;

*heterodont hinge:* teeth differentiated, perpendicular to the cardinal margin, also lateral teeth, parallel to the cardinal margin;

*disodont hinge:* with few teeth, or none at all;

*schizodont hinge:* fewer cardinal teeth, strong and grooved; one single tooth separating two recesses on the left valve, and two teeth separated by one recess on the right valve;

*isodont hinge:* recesses and teeth symmetrical in relation to the axis of the shell;

*desmodont hinge:* similar to the heterodont but with projections and indentations due to the action of the ligament;

*pachiodont hinge:* characteristic of the rudimentary types now extinct; few teeth but large and deformed.

The last five types are principally no more than modifications of the heterodont hinge.

The first lamellibranchs appeared during the Cambrian period. Over the following eras, large numbers of new groups appeared, some of which became extinct while others continued virtually unchanged to the present day. As happens with other invertebrates, the evolution of this class is difficult to reconstruct, both because of the wide variety of animals included in the group, and because of the lack of data about them.

Why, then, are the lamellibranchs important for palaeontology? The reason is simple: when

they are found in a rock they enable one to establish the conditions in which the rock was formed and consequently permit palaeographic maps and charts to be drawn of the period in which fossils and rock were deposited. In addition to this, there are a number of lamellibranchs that serve as guide fossils for certain geological periods.

From the point of view of classification, the lamellibranchs are divided into three large orders, based upon the shape of the hinge, the musculature and the type of impressions, this last being used after comparisons between the fossil forms and the corresponding living forms.

The taxodonts (Taxodonta) of which there are a large number of genera still alive today, include the oldest and most primitive lamellibranchs, which typically bear a taxodont hinge and two muscular impressions on each valve. It is to this group that the genus *Ctenodonta*, found in Cambrian (or possibly Ordovician) rocks in Portugal, is attributed. It is the oldest lamellibranch known. This fascinating order also includes some of the most common of the lamellibranchs, the first freshwater forms from the Carboniferous and Permian periods.

The order Anisomyaria includes those lamellibranchs in which the anterior muscle is reduced

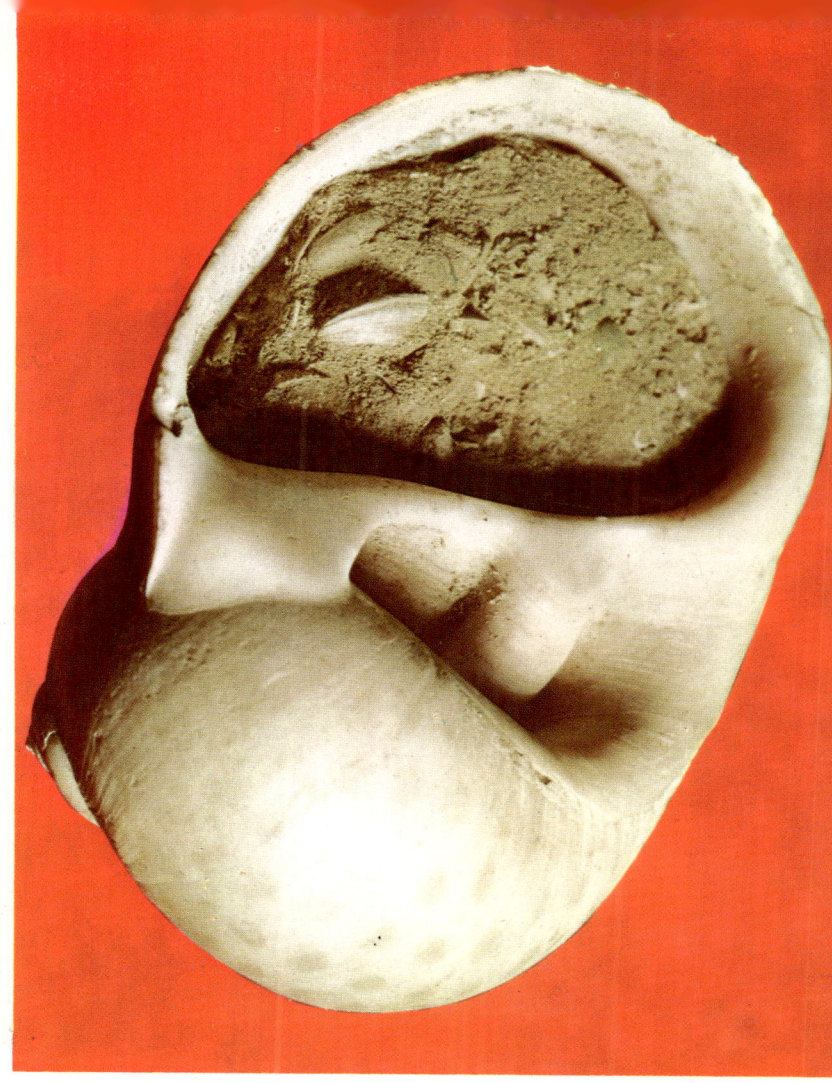

or lacking and the posterior muscle, which is well developed, is displaced towards the centre of the valve. They have no siphon, and the hinge is disodont or isodont. Included in this group, which has been known from the Ordovician period down to the present day, are the pettinides, ostreides and spondyli, species of which have provided large numbers of important guide fossils over the years.

Finally, the order Eulamellibranchia includes all the other lamellibranchiata known, and is thus by far the widest and most important subdivision. The eulamellibranchia have equal anterior and posterior muscles, schizodont, heterodont or desmodont hinges. Also included is the extinct group known as the rudistes, lamellibranchiata with a pachiodont hinge which lived from the Upper Jurassic period to the end of the Cretaceous and modified very slowly over the years. The rudistes attached themselves to the substratum by means of one of the valves, which became much larger than the other, assuming the shape of a twisted horn of an almost straight cone. The other valve was transformed into a type of operculum that was hinged to the first by means of large deformed teeth and recesses. Among the most common types, which are used as guide fossils to date reef formations of these remote periods, we should mention the genus *Requienia* of the Lower Cretaceous period, which has twisted and pointed valves, with a groove running between them.

Within the family Hippuritidae are grouped a number of slightly different genera, among them the well-known *Hippurites*.

All the animals of this family had a conical right valve that was attached to the substratum by means of its apex and the left opercular valve. On the larger valve there were three longitudinal furrows corresponding to internal folds of the shell, the first of them called the bonding furrow, the others, pilasters. At the end of the Cretaceous period the rudistes became extinct. Their lesson to us, when compared with the corals and the reef brachiopods of that era, is that animals of different phyla can, in the same environment, become very similar in outward appearance.

## Gastropoda

The gastropods are a vast group of molluscs which, unlike the lamellibranchs, include some species that are adapted to subaerial surroundings. The class comprises molluscs that are characterized by a spiral univalve shell, together with some species that do not possess a shell, but which nevertheless exhibit torsion of the viscera

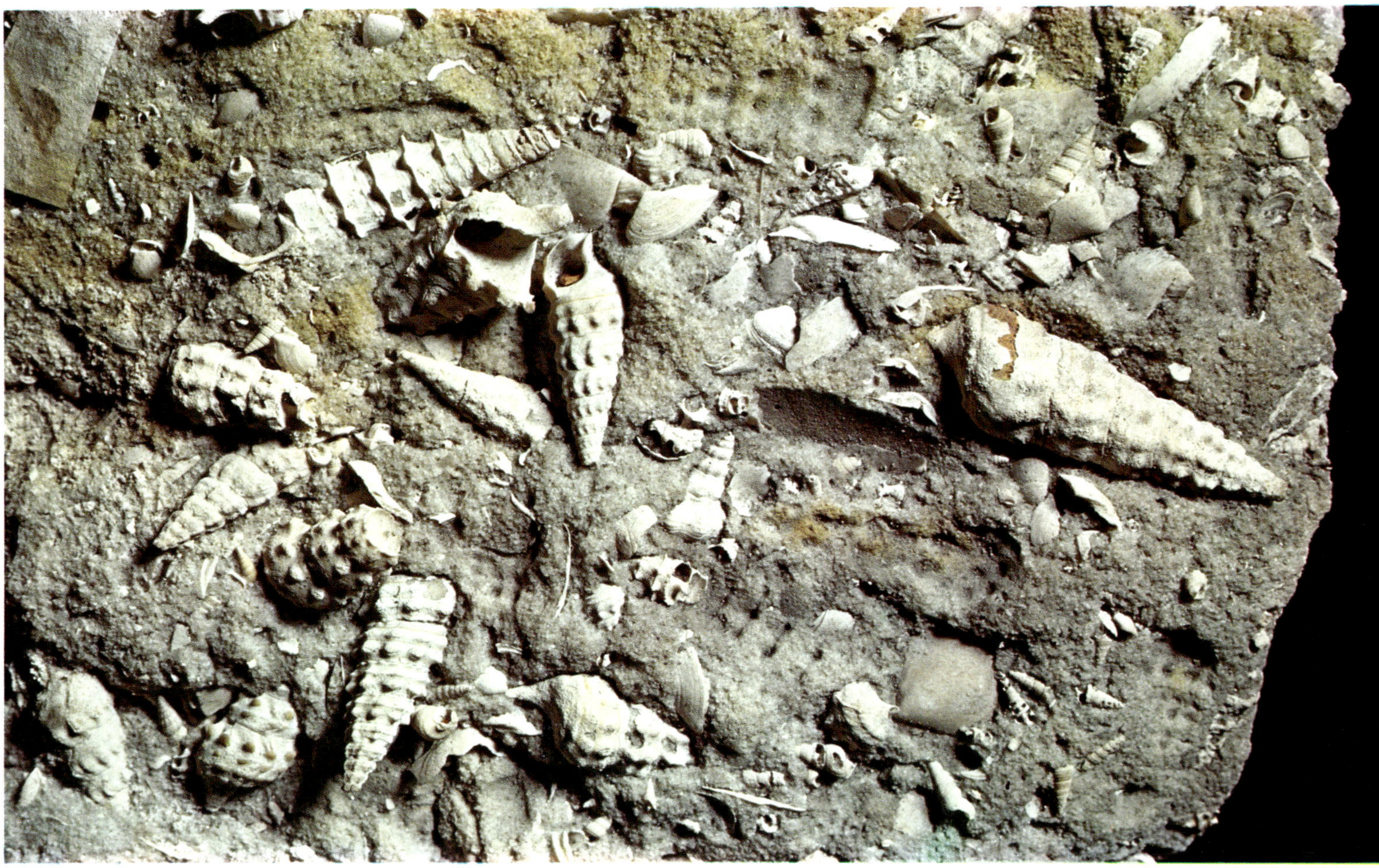

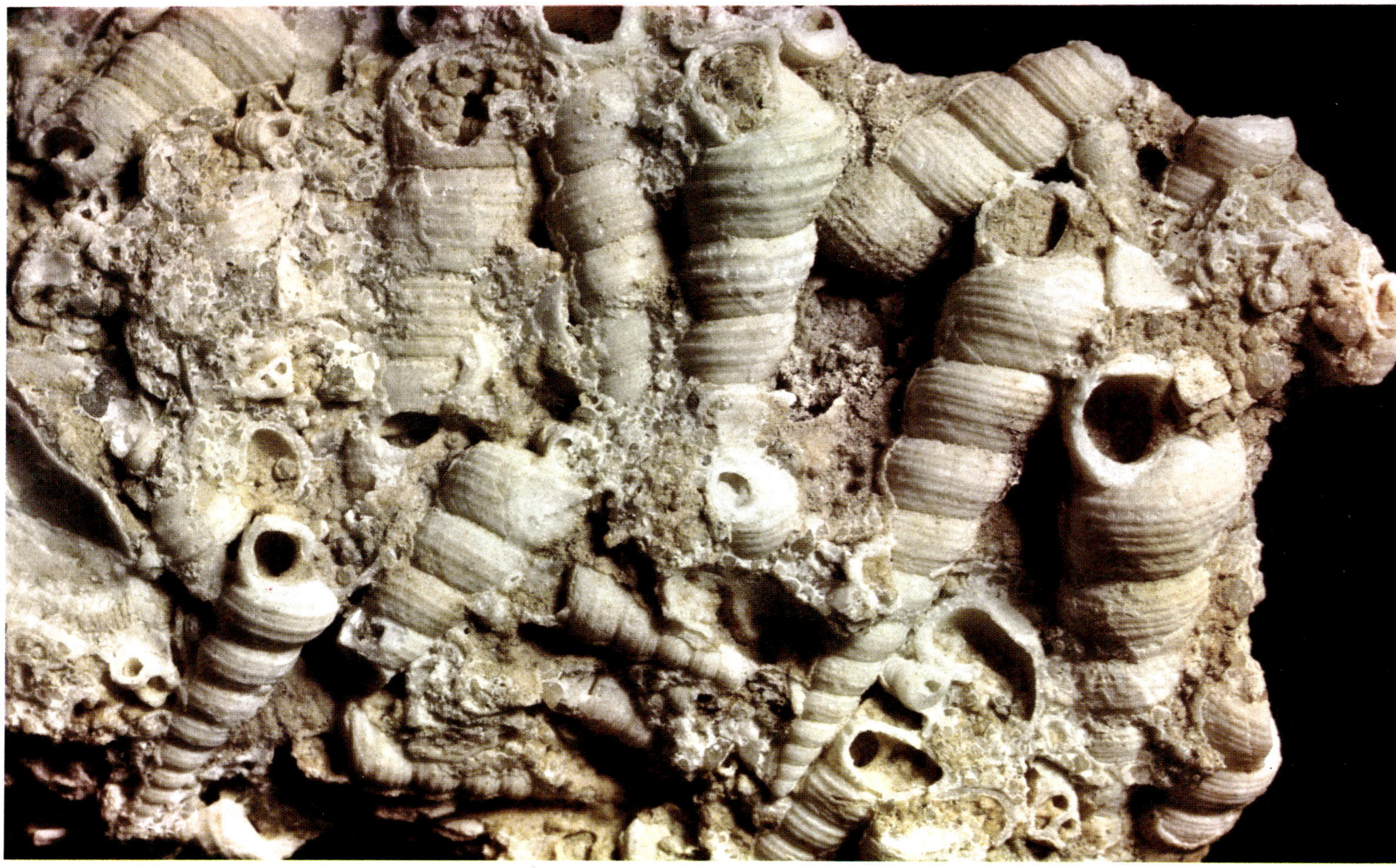

The shell of a gastropod, which, as in the other groups, is the only part that is preserved in the fossil state, can be thought of as an elongated cone wound into a spiral. The direction in which the spiral winds is generally right-handed (the shell pointing upwards); a left-hand spiral may be due to malformation or it may be a characteristic of the species. The successive turns of the shell are detached from each other, and more or less overlap; in some cases the last turn, which ends in the aperture, embraces all the others – as happens, for instance, in the cowrie shell.

In a typical shell, the point is known as the apex or the posterior end, and all the coils except the last one are called the spiral. The angle measured at the apex is known as the spiral angle. The line that separates the various turns of the spiral is called the suture line and has a very variable appearance. In species with serrated turns, a type of axis (columella) is produced inside the shell; if however, the turns are broad at the base this is replaced by a cavity (umbilicus) which is sometimes blocked by a callus, itself an excellent aid to classification.

The aperture of the shell, also useful for classification, is very different from species to species. Regularly circular in the primitive types, it becomes oval, elongated and fairly irregular in the evolved types. Its edge (peristoma) incorporates an inside edge (columella) and an outside edge (lip). The peristoma can be complete, in which case the animal is olostomic, or intagliated, and the animal is then said to be siphonostomic. In such animals, two crevices may exist, one frontal or branchial and one posterior or anal. These serve for the passage of the animal's two siphons. In forms without a siphon the lip sometimes has a groove in the middle and this is the means by which water passes into the branchial cavity. This fissure however becomes progressively cemented up during the growth of the shell, leaving a characteristic furrow on its surface (pleurothomarinal suture). This is generally to be seen in the less evolved forms, since the siphon appeared for the first time only at a late stage in the history of the gastropods, and thus signifies a more advanced animal.

When the animal withdraws into the shell, the aperture can be blocked by a mobile calcareous or horny part, carried by the posterior dorsal part of the foot, and called the operculum. This is rarely preserved.

The decoration of the shell, which is extremely important in the classification of the fossil forms, is extremely varied: the shell can in fact be completely smooth; it can be furrowed by light growth lines or by well-defined ridges; it can have large

(counterclockwise twisting of the soft organs of the animal), with a consequent loss of symmetry. They are much more mobile animals than the lamellibranchs, since they can move by means of the disc-shaped muscular foot situated outside the sheath in a ventral position. There are species that attach themselves to the bottom.

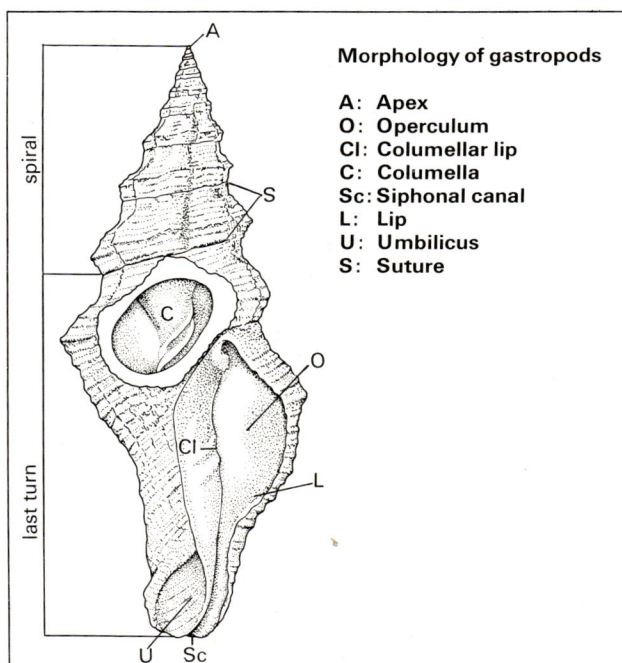

**Morphology of gastropods**

A: Apex
O: Operculum
Cl: Columellar lip
C: Columella
Sc: Siphonal canal
L: Lip
U: Umbilicus
S: Suture

folds, spines, tubercles and varices along the whole height of the turn. There are two principal types of ornamentation: axial ornamentation, in which these elements run parallel to the axis of the shell, and spiral ornamentation, when they are parallel with the development of the spiral. From an examination of the fossil forms it can be seen that the ornamentation has become steadily more complex with the passage of time. One mysterious fossil group, the nerines, which lived a reef existence, carry a particular decoration inside the shell instead of on the external surface.

The classification now adopted for the gastropods is principally based upon the anatomical features of the soft body of the animal; it is thus a strictly zoological classification, followed by palaeontologists because fossil species bear a remarkable similarity to present-day forms. The class Gastropoda is further subdivided into three subclasses: opisthobranchiates (Opisthobranchia), prosobranchiates (Prosobranchia), and pulmonates (Pulmonata), classified on the basis of the position of the branchia in relation to the heart in the case of the first two, and on the presence of a sort of pseudo-lung in the third, which groups together freshwater and terrestrial gastropods.

The first true gastropods appeared in the Lower Cambrian era, about 500 million years ago. Bilaterally symmetrical molluscs, with a flat spiral shell, in which the soft parts were only slightly twisted, they belonged to the order Archeogastropoda, which was to be found throughout the Palaeozoic period. The first opisthobranchiates were, however, found in the Devonian period, and have increased in size and number over the succeeding millennia. The last gastropods to appear were the pulmonates, which were first seen during the Carboniferous period. By the beginning of the Cenozoic era, about 60 million years ago, the gastropods had begun to look very much like present-day forms, and have not changed very much since that time. (Incidentally, it is interesting to find that the lamellibranchs were much more common than the gastropods during the Palaeozoic and Mesozoic periods, while from the Tertiary period and until the present day this situation has been reversed.)

While these molluscs are not as important for dating terrestrial rocks as the groups that follow, nevertheless they are extremely useful, even more so than the lamellibranchiates, in the reconstruction of the conditions of the past. In fact, knowing the present way of life of such animals and assuming a similar pattern for the fossil species, we are able to draw conclusions about the appearance of the sea bottom and its

turn divided into three distinct orders, Archaeogastropoda, Mesogastropoda and Neogastropoda, all of which are characterized by systematized branchia in front of the heart.

The most primitive gastropods lacking a siphon, and with a pleurotomarian suture running through them, belong to the first of these. The shell form is very varied; some are found with a flat spiral shape, like the genus *Bellerophon*, which existed from the Ordovician to the Permian periods, and is very common in some alpine Palaeozoic layers. Some have a regular helical shell, some are patelliform, like the genus *Fissurella* in which the pleurotomarian suture is reduced to a hole located at the apex, while yet others are shaped like an ear. Among these last is the genus *Haliotis*, which appeared during the Cretaceous period and is still alive today, and in which a series of aligned apertures indicate the

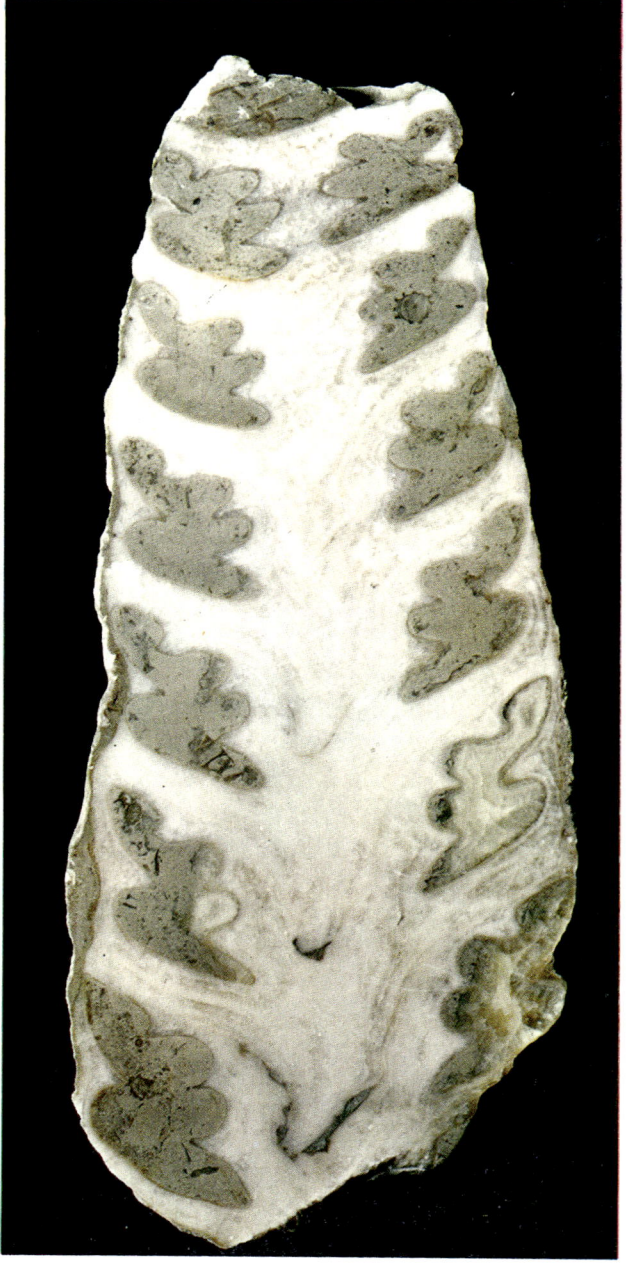

depth. Thus, for example, the genus *Nassa* indicates a muddy bottom, *Natica* and *Turritella* a sandy bottom, and *Patella* a rocky bottom, while the genera *Patella*, *Fissurella* and *Haliotis* indicate coastal waters, and *Helix* a subaerial location. Similar useful information is obtained when such organisms are considered as climatic indicators — one has only to look at present-day fauna to see that molluscs typical of warm seas and of cold seas have totally different characteristics, and similar differences are to be found among fossils.

### —Subclass Prosobranchiata

Turning then to the classification of the gastropods, the majority of fossil genera belong to the first subclass, the Prosobranchiata, which is in

*Top:* Xenophora crispa, *Pliocene; Modena Appennines. Height of the original, 3.6 cm. Lower:* Vermetus *sp., a gastropod with an irregular shell, encrusting the valve of an oyster. Pliocene; Asti, Italy (lifesize). Right: Section of a specimen of* Nerinea *sp. The characteristic folds of the internal ornamentation can be seen. Upper Cretaceous; Austria. Height of the original, 10 cm.*

*Right: Some sectioned specimens of* Acteonella *sp. Upper Cretaceous; Brandenberg, Austria. Below: Conical shells of* Tentaculites *sp., a genus included in the opistobranch gastropods. Lower Devonian; Schieferngebirge, Germany (much enlarged)*

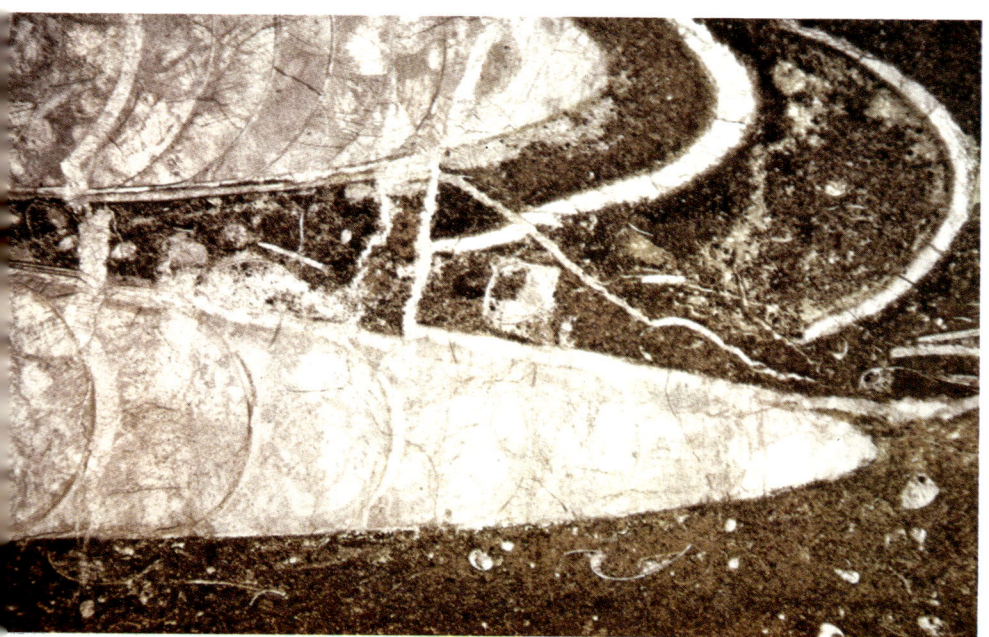

path of the incompletely blocked suture.

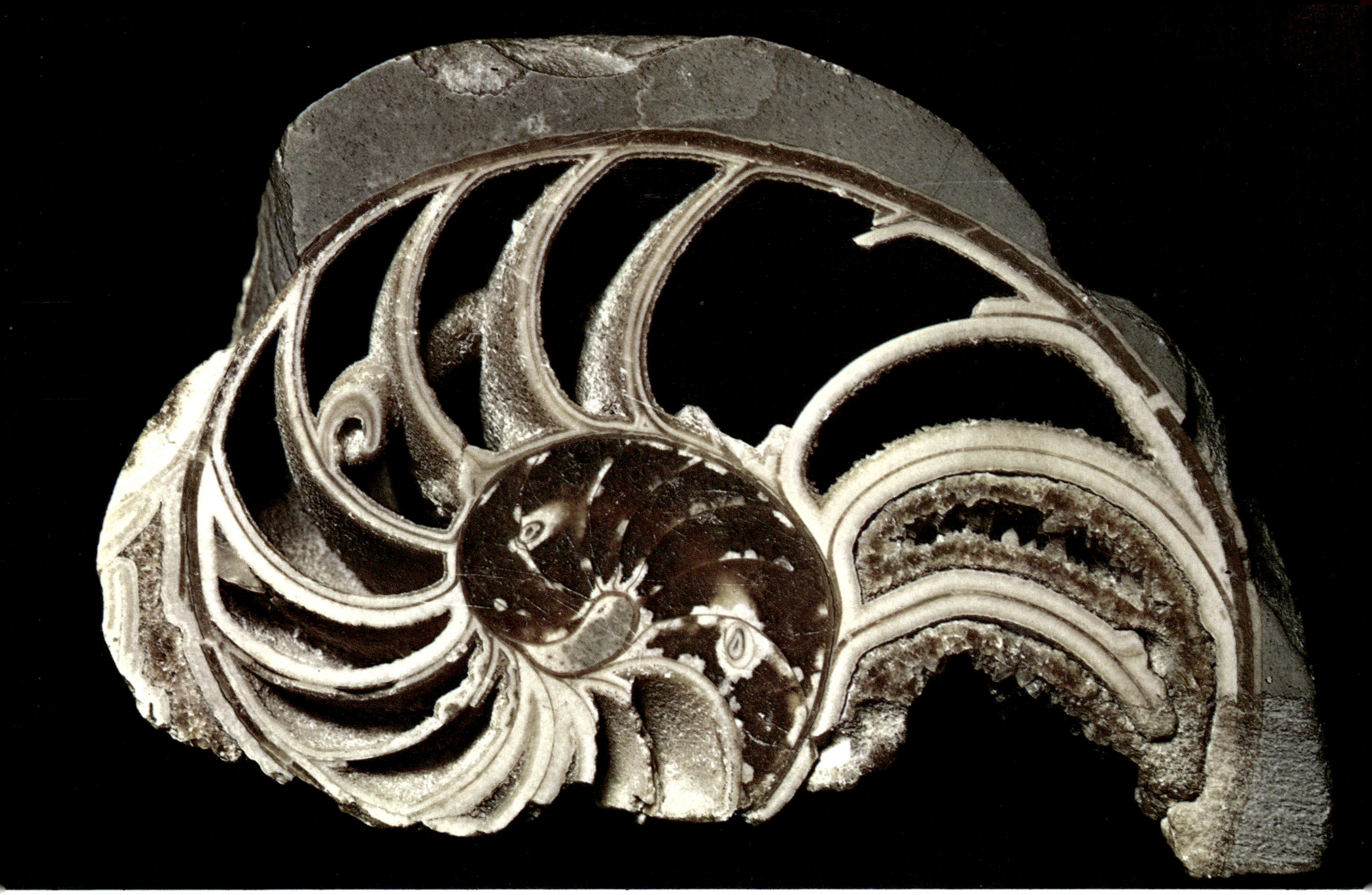

Some genera that are well known to the collectors are attributed to the order Mesogastropoda, which appeared a little later in the Lower Ordovician period. Modern representatives of this order possess the most beautifully coloured shells, which unfortunately are not preserved by fossilization. They are oceanic, freshwater and terrestrial gastropods with for

*Below: Thin section in limestone of* Orthoceras. *The right-handed, chambered shell is typical of the primitive nautiloids. Silurian; Flumini-maggiore, Sardinia (×9)*

the most part helical shells, or – more rarely – shells in the form of a cowl or discus. These animals, too, lack a siphon, and are grouped together on the basis of complex details of their internal organs. Among these are the fossil nerines, which lived in the Jurassic and Cretaceous periods in the vicinity of reefs, and whose shell has a particular internal decoration that can be seen in the sectioned specimens.

Finally, gastropods that appeared in the Lower Ordovician bearing a siphonal canal are attributed to the order Neogastropoda. Many members of this order, such as murices, whelks, snails and cones, which were widely distributed in the Tertiary era, are still to be found today. Fossils of these animals are found in marine sedimentary rock, but lack the colours possessed when alive.

### —Subclass Opisthobranchiata

The second subclass, the Opisthobranchiata, comprises some marine gastropods that have a small simple shell and the branchia behind the heart. They are known in the fossil state from the Carboniferous period; only the pteropods, oceanic molluscs provided with a conical shell in which the foot forms two lobes or paddles used for swimming, seem to have appeared earlier, in the Lower Cambrian period. As occurred in the past, so today, on the bottom of the open seas at depths of between 3,000 and 8,000 feet, sediments entirely made up of accumulations of the shells of these organisms are formed.

Some strange now-extinct organisms, of which only the conical shell is known, are very similar to the pteropods; these are grouped in the genera *Tentaculites* of the Ordovician and Devonian periods, and *Hyolithes* of the Cambrian and Permian periods.

The best-known genus belonging to this sub-class, however, is the globular-shelled *Acteonella*, characteristic of reef surroundings of the Cretaceous period, of which it is a guide fossil. And finally, the nudibranchs, which have either small internal shells or no shells at all, and consequently are virtually unknown in the fossil state, are attributed to the opisthobranchiates. A single family of them survives from the Eocene period.

### —Subclass Pulmonata

The last group of gastropods in our classification is that of the pulmonates, which lack branchia, but possess a simulated lung or respiratory sac, formed through the fusion of the sheath and the viscera. They are the second largest group of gastropods, comprising more than 7,000 species, of which 6,300 are living, and about 700 are fossils. Members of the pulmonates are terrestrial gastropods, such as snails and cochlea, and include many freshwater gastropods.

The first pulmonates appeared in the Palaeozoic period, but attained their present-day distribution and importance only during the Upper Cretaceous period. They are classified in two orders: Basommatophora, whose members have eyes at the base of their head tentacles, and Stylommatophora, whose eyes are situated at the top. Of the basommatophores, which are all freshwater species, the genus *Planorbis*, which first appeared in the Jurassic period and is widespread in Tertiary sediments and in present-day fresh waters, is particularly notable. The stylommatophores include the most common snail, the genus *Helix*, which appeared in the Cretaceous period and is frequently found in the subaerial sediments, such as those deposited in the Tertiary and Quaternary fossil dunes.

### Cephalopoda

The cephalopods are a very old group of highly specialized molluscs. The 400 cephalopod species living today are in fact only a small fraction of those living for 500 million years from the Upper Cambrian period. (The few species belonging to

the genus *Nautilus* are, for example, what remains of a vast group to be found in all seas for several million years.) The cephalopods have always been marine animals, and from the morphological and anatomical point of view, are the most evolved molluscs of all. They are found today in all seas, and their fossils, also widely distributed in sedimentary rocks, are useful as guide fossils.

The name 'cephalopod' comes from the fact that the foot – which, as we have already seen, has various forms and functions in the different mollusc classes – is partly transformed in these animals into a series of tentacles that surround the mouth, and used both for locomotion and in the capture of prey. Another part of the foot forms the siphon, an organ that the animal used to move itself by 'jet propulsion' – that is, by expelling powerful jets of water. Unlike the earlier groups, the head of the cephalopod is quite distinct and bears two large lateral eyes. Horny mandibles are found in the mouth, and these fossilize easily. The principal characteristic is the presence in almost all of these animals of an external or internal shell which also fossilizes easily.

The shell is different from those of other molluscs in that it possesses numerous sectors that divide it internally into chambers. It is on the basis of this feature that certain fossil forms now

*Right: A perfect ammonite from the English Jurassic, with the lobal lines clearly marked along the whole development of the spiral. Diameter, 13 cm. Left: Aturia aturi, a Miocene nautiloid with characteristic zig-zag suture pattern. Alexandria. Diameter of the shell, 4 cm*

extinct have been attributed to the cephalopods. The most important of these are the ammonites, which lack soft parts and could not otherwise have been classified.

The class Cephalopoda is divided by palaeontologists into the subclasses of the nautiloids (Nautiloidea), members of which were very common during the Palaeozoic period and of

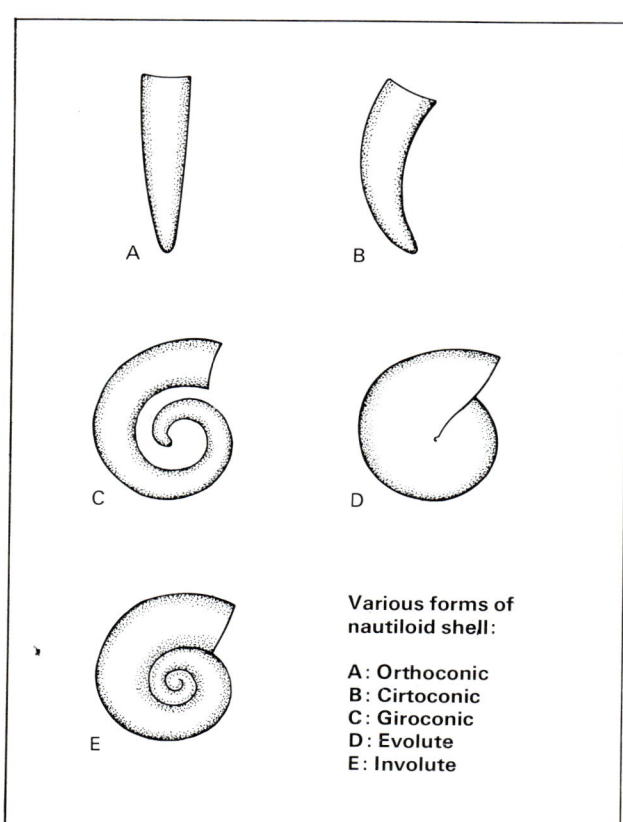

Various forms of nautiloid shell:

A: Orthoconic
B: Cirtoconic
C: Giroconic
D: Evolute
E: Involute

*Right:* Cenoceras striatum, *a nautiloid of the Lower Jurassic, very similar to the surviving modern species. Alpe Turati, Como, Italy. Diameter of the shell, 10.5 cm*

which the genus *Nautilus* is the only one still living, the ammonoids (Ammonoidea), which consist entirely of fossils that lived from the Devonian to the Upper Cretaceous period, and coleoids (Coleoida), which include the present-day cephalopods, such as the cuttle fish, which are very similar to fossil forms. These subclasses will be considered individually.

### —Subclass Nautiloidea

The nautiloids are the oldest cephalopods; they appeared suddenly in the Upper Cambrian period, and little is known of their relationship to other groups of invertebrates. But although their origin is obscure, these are nevertheless an extremely important group since they are thought to have been themselves the origin of the

*Above: Section of an ammonite in the pink 'marble' of Verona, showing notable chambering of the phragmocone. Upper Jurassic. Diameter of the original, 11.4 cm*

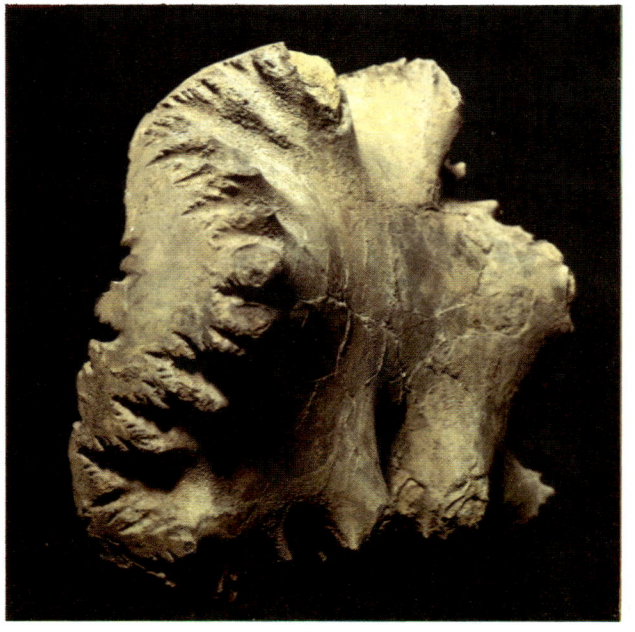

subclasses of the ammonoids and the coleoids. Moreover, unlike the other extinct groups, the nautiloids have a few links with present-day species, after a life that lasted over 500 million years. Indeed, there are still some species of the genus *Nautilus* to be found in the seas of today, principally on the western side of the Pacific Ocean. Observation of the older nautiloids, however, shows that these have changed with the passage of time and that present-day species are not identical with the older types. In the course of evolution there have been remarkable developments in the form of the shell, the internal anatomy, and the habitat and way of life.

The first Cambrian nautiloids were thus very primitive types, with a straight or slightly curved shell. During the subsequent periods of the Palaeozoic era an explosive evolution took place; the species suddenly began to increase in number and the shells tended to have a flat spiral twist. This twisted form was arrived at as early as the beginning of the Silurian period, when the development of the nautiloids was complete. Intermediate stages in their development display a cyrtoconic, or slightly arched shell, a gyroconic shell (twisted without a suture between the successive turns), an evolute shell with turns that hardly touch, or the involute shell in which the

last turn overlaps the preceding turns. This final stage is to be seen in the modern *Nautilus*, whose last turn embraces all the preceding turns.

At the end of the Palaeozoic period the nautiloids began to diminish in numbers, probably as a result of the ever-increasing numbers of ammonites which, as we have said, developed from them, and began to appear around this time. The diminution in numbers continued progressively throughout the succeeding epochs until reduced to the single genus that is still to be found today, 150 million years later.

The presence of an external calcareous shell is the feature that is of most interest in the study of these animals. It tends to be divided into two main parts, the phragmocone and the living chamber. The phragmocone is the more developed part of the shell, and is divided internally into chambers by septa that are perpendicular to the longitudinal axis of the shell and have a concavity turned towards the aperture. These septa produce characteristic insertion or suture lines on the surface of the shell. In each septum there is a hole through which runs the siphon, a soft organ that passes from the body of the animal through the various chambers up to the embryonal gallery, which is located at the centre of the spiral. The chambers are called air

chambers, and contain a gaseous mixture similar to atmospheric air but richer in nitrogen, and this supply is controlled by the animal to make the shell lighter or heavier, so that vertical movements can be carried out.

The living chamber is the inhabited part of the shell, bounded behind by the last septum and open towards the outside in a way that varies according to the type of animal. The aperture may be narrowed by lateral formations on its edge (the peristome); this often gives rise to curious apertures, such as the T-shaped aperture to be seen in some genera of the Silurian and the Carboniferous periods. On the outer surface of the shell decorations may be seen, formed by simple growth lines, or by parallel or longitudinal ridges and furrows. In some rare fossilized specimens, under certain conditions of preservation, it is possible to make out the external colouring of the shell, formed by light or dark longitudinal or transverse grey bands and lines.

The subclass Nautiloidea is further subdivided by palaeontologists into 14 distinct orders, the majority of which existed during the Palaeozoic period. Among the oldest and most commonly found are the orthoceratides, which have a straight shell and a habitation chamber that can attain a length of six feet in the giant forms, and

are decorated by longitudinal or transverse ridges. Members of this genus lived from the Ordovician to the Triassic period. The family Nautilidae appeared during the Upper Triassic period; and it is to this family that the genus *Nautilus*, which we have already mentioned, belongs.

The study of this one living genus furnishes us with valuable clues about the way of life of the fossil nautiloids. *Nautilus* lives in warm seas, and apparently its habitat is closely linked with both the temperature and the degree of salinity of the water. Its usual habitat is at a depth of 1,600 feet, but it moves to the surface each night, probably following the movements of the plankton on which it feeds. It is an excellent swimmer and its hollow shell floats and may be transported a long way by the current, which is the reason why the nautiloids are so widely distributed, and hence why they are so useful as guide fossils.

### —Subclass Ammonoidea

The ammonites are one of the great enigmas of palaeontology. They are an immense group which was widespread in the seas of the Mesozoic period, but then became completely extinct at the end of the Cretaceous period without any obvious explanation for such a decline. Their external shell, similar to that of the nautiloids, fossilizes

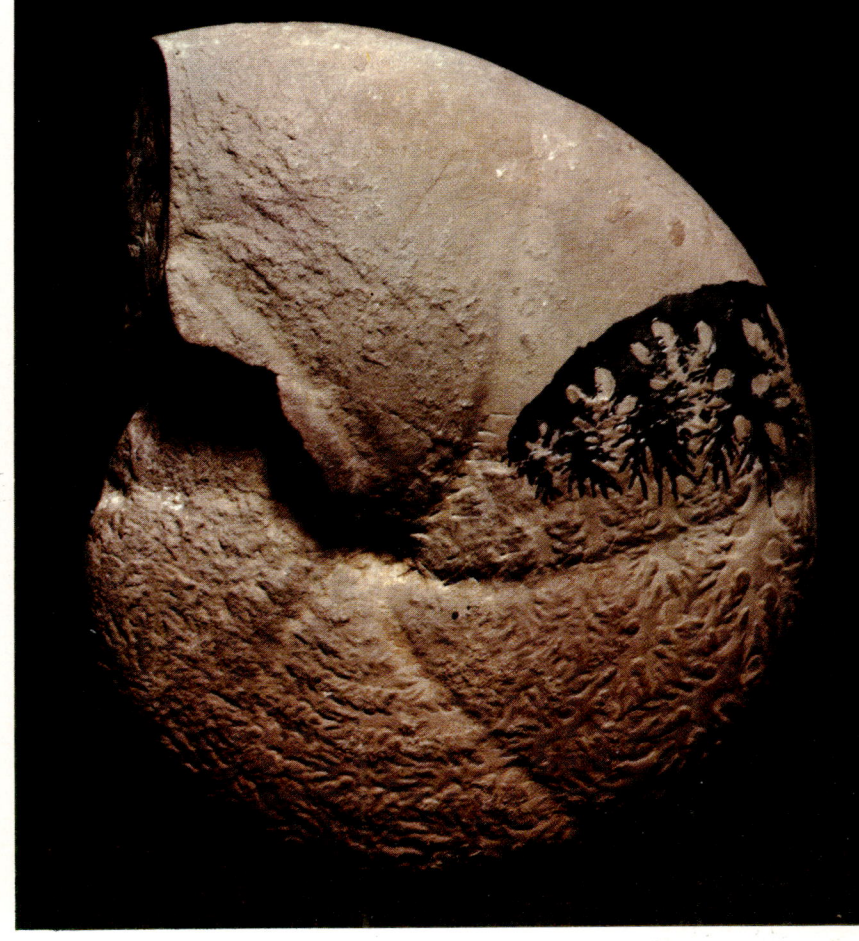

perfectly because of its solid calcareous shell, and is, in fact, the only part that we know. The soft organs of the body are completely unknown, but are thought to be similar to those of the present-day *Nautilus*.

An ammonite shell is made up of a single valve, generally a flat spiral, as if a long and narrow cone had been twisted about its axis. It is divided into three parts: the protocone or initial chamber, which forms at the beginning of the animal's life, and represents the embryonal stage of the ammonite; the phragmocone, a long compartmented part divided into numerous chambers by septa and probably filled, as in *Nautilus* species, by a mixture of gases; and finally the living chamber, which opens on the outside of the shell, in which the animal lived. It was able to withdraw itself completely inside its shell and close the aperture with a calcareous operculum that was situated on the ventral part of the sheath. The animal and shell were joined by the siphuncle, a membrane impregnated with calcium phosphate which passed out of the posterior end of the body, through the various chambers by means of holes in the septa, and penetrated the protocone with an enlargement (caecum) joined at the bottom to a calcareous cord – the prosiphon.

The phragmocone, the part that fossilized best, constituted the more developed part of

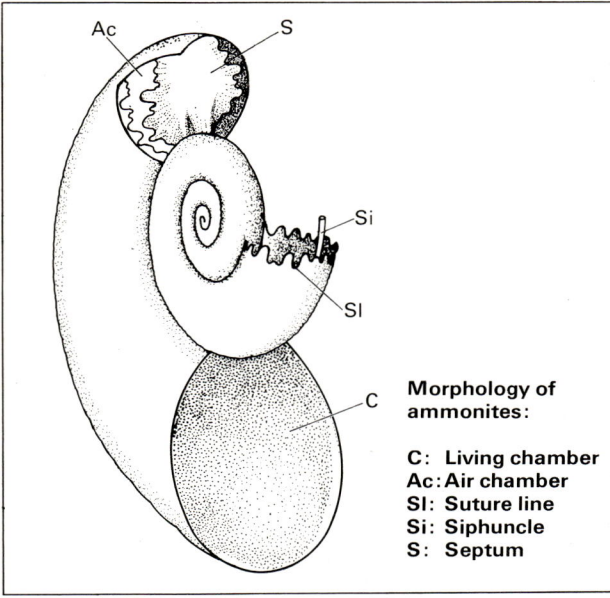

Morphology of ammonites:

C: Living chamber
Ac: Air chamber
Sl: Suture line
Si: Siphuncle
S: Septum

first ammonites, which appeared in the middle of the Palaeozoic period, were straight, and that the shells did not become twisted until much later in the Lower Devonian period, and by the Cretaceous period the animals display flat spiral shells with tightly wrapped turns. The Cretaceous period, however, also marks the beginning of the decline of the ammonites, for they began to develop strange forms that were twisted in an irregular way: species that were straight, hooked, partly twisted, and partly straight, curled up in an odd way and so on. And so the entire group began to die out after a life of 250 million years (from the Lower Devonian period to the end of the Cretaceous), finally disappearing about 60 million years ago, leaving only fossilized remains as evidence of their existence.

Even today, very little is known about the life of the ammonites. Two principal obstacles impede work on these animals: lack of soft parts and the probable post-mortem transportation of the shells. This latter possibility in particular makes it very difficult to establish a relationship between the fossil and the conditions in which it was deposited. And such a transportation was indeed very common, since the shell of the ammonites, like that of the nautiloids, was very light, and its chambers full of air. Once the body

the shell, and was divided into chambers by complex septa that left a characteristic insertion line, known as the suture line, on the surface of the shell. Finally the terminal part of the shell consisted of a large chamber occupying between one and one and a half turns of the spiral. Its external edge (peristoma) had a variable shape.

Ammonite shells are most often found to be covered by a very pronounced decoration made up of radial, straight, curved and sinuous ridges, by tubercles of different sizes, and by spines that could attain remarkable dimensions. The subdivisions into genera and species are in fact based upon this ornamentation, while for subdivision into much larger groups such as the suborders and families it is the suture line that is more important. This line, as we have said, is produced on the external shell of the ammonite by the various septa of the phragmocone, and since these septa had no linear progression, it is frequently broken by sinuosities called 'lobes', and large protuberances known as 'saddles'. In the course of the animal's evolution, the suture line became progressively more complex, passing from simple forms to extremely complicated outlines. The same development is also, in fact, to be seen during the growth of each individual animal, so that in the young ammonite the line is very simple, tending to become progressively more complicated during growth. It is believed, even if still not proven, that this development of the septa served to strengthen the thin shell, thus making it more resistant to injury and pressure at great depths.

Another feature used in the classification of these fossils is the shape of the shell which, as in the case of the nautiloids, can vary enormously from one animal to another. It seems that the

*Above left:* Harpoceras exaratum. *Lower Jurassic; Turati Alps, Como, Italy. Diameter, 6.1 cm. Above right:* Leioceras opalinum. *Middle Jurassic; Germany. Diameter, 4.2 cm*

had gone it could have floated a long distance.

The fact, however, that there were so many types of ammonites, each with its different shape and decoration, leads us to suppose that the ammonites were extremely adaptable, and could be linked to particular surroundings. Studies have in fact shown that disc-shaped or flat shells are better adapted than others to swimming, and that they are consequently more frequently fossilized in clay and marl, rocks that are mostly indicative of deep sea. Highly decorated shells, on the other hand, are found solely in calcareous rocks, which indicates deposition much closer to the coastline. Appreciable results have been obtained just by studying swimming, and it has for instance been possible to establish, by experiments with models, that the ammonites moved about holding the shell with the twisted plane vertical and the aperture towards the sea bottom. This seems to be confirmed by the discovery, in the Jurassic of Solnhofen, of fossilized ammonite shells beside the imprint that was left when they fell on to the sea bottom. Other clues as to how these animals lived are furnished by the study of other animals of the same era. For example, despite their hard shell, these animals were an important food for contemporary marine reptiles. Many examples have in fact been found in the stomach of these fossilized animals, and one large and particularly well-known ammonite shows signs of having been bitten by a mosasaurus, a giant marine reptile of the Upper Cretaceous period.

The subclass Ammonoidea comprises the single order of Ammonitida, which is in turn divided into numerous suborders: Anarceatina, Goniatitina, Certitina, Phylloceratina, Lytocertatina, Ammonitina; each of these comprises in its turn numerous superfamilies, families and genera. The importance of the ammonites to palaeontologists is principally based on the fact that they are used as guide fossils. Each species is in fact characterized by a very wide distribution over an extremely short period of time. Thus on the basis of the types of ammonites found it has been possible to divide each part of the Mesozoic era into numerous 'palaeontological zones', each corresponding to a period of time in which one or more characteristic ammonite species are present; these zones may be linked accurately.

We must add a few words on the aptychi, calcareous ammonite opercula that are found isolated in the sediments. Rare examples of ammonites with the aptychi in the original position have made possible the classification of these strange fossils.

The aptychi can be divided into two large categories: the anaptychi, widespread in the sedimentary rocks dating from the Upper Devonian to the Cretaceous period, which consist of a single 'valve' decorated with concentric and radial ridges; and the true aptychi, limited to Jurassic and Cretaceous terrains, and formed by two boldly ornamented 'valves'.

Because they are found by themselves, it is only rarely possible to determine the genus of ammonite to which different aptychi belong.

### —Subclass Coleoidea

The subclass of the coleoidea brings together the majority of the cephalopods alive today. In the past geological eras they were well represented, without however reaching the distribution attained by the nautiloids and the ammonites.

The coleoidea possess only two branchia, two less than the nautiloids, and, probably, than the ammonites. They are animals known to all, since they include the squids and the cuttlefish, which are found all over the world.

Leaving aside the anatomy of the soft parts, we should however mention that the majority of these animals possess an internal shell (eg the 'bone' of the cuttlefish), and this is usually the only part of the animal that is fossilized. Very few species possess an external shell. In fact the fossil remains of those coleoidea, such as octopuses, that do not possess any shell are extremely rare.

Although internal and very different from that of the other cephalopods, the shell nevertheless has the same general structure and can be divided into three parts, roughly corresponding to those of the other two subclasses.

The coleoidea includes two orders. The first has 10 tentacles around the mouth, and this order is in its turn divided into one extinct suborder, the belemnoids (Belemnoidea), and two living suborders, the sepioids (Sepioidea) and the teuthoids (Teuthoidea). The second order comprises the octopuses (Octopods), which have eight tentacles, and which include the extinct suborder of the palaeoctopods (Palaeoctopoda), and the polypodoids (Polypodoidea), which have numerous representatives, both fossil and living.

In the Mesozoic sediments of marine origin, strange elongated fossil remains in the shape of a cigar are often found. These are extremely strong and completely calcified, and for some time have constituted a real enigma for palaeontologists. Detailed studies of their internal anatomy, as well

*Below left:* Parapeltoceras annulare. *Middle Jurassic; Germany. Diameter, 4.8 cm. Below right:* Rasenia uralensis. *Upper Jurassic; Britain. Diameter, 2.6 cm*

known as a mucrone, or is sometimes rounded. The surface is frequently covered by granulations and furrows, and these last, which developed principally in the ventral and dorsal parts, are important features in classification, since they are considered to be traces of muscular insertions. In the anterior part the rostrum has a conical cavity, within which the phragmocone is located. This is the septate part of the shell and corresponds to the compartments of the nautiloids and of the ammonites. Because of its extreme fragility it is very rarely preserved. It is formed by a certain number of horizontal septa, crossed near the ventral edge by a siphon. In the best-preserved specimens it is possible to see the very delicate external walls of the phragmocone, while the proostracum is even more delicate and hard to find.

The first belemnites appeared at the beginning of the Jurassic period. These first belemnites had a shell with a phragmocone that was relatively much larger than the rostrum, which was very reduced in size and lacked a proostracum. In the following geological eras the phragmocone decreased in size in relation to the rostrum, which then in turn became smaller than the proostracum. Thus in the aulacoceratides, the most primitive belemnites of the Jurassic period, the phragmocone is much more developed than the rostrum and the proostracum is lacking;

*Left:* Olcostephanus astierianus. *Lower Cretaceous; Germany. Diameter, 3.2 cm. Below:* Epihoplites denarius. *Lower Cretaceous; Folkestone, Britain. Diameter, 3.5 cm*

as of the few impressions that have been found of their soft parts, have finally made it possible to establish that the tough 'cigars' are all that remains of the internal shell of some extinct cephalopods, which were called belemnites, and grouped in the suborder of the belemnoids. Unlike the ammonites, whose soft parts are unknown to us, we have, as has been said, records of the imprints of the body of some belemnites, left by the animal falling after death on to the thin layer of mud that covers the bottom of calm marine basins. From reconstructions made on the basis of these recordings, it is thought that the animal was very similar to the cuttlefish: it had an ink sac and a number of tentacles, which seem to have varied from eight to six and were situated in the cephalic part around the mouth, which had small mandibles.

Inside the animal's body was the shell, composed of three parts: the rostrum, or guard, the phragmocone, and the proostracum. The rostrum is the strongest part, and it is this that is preserved best in the fossil state. It is made up of small prisms of calcite disposed radially around the longitudinal axis, and is generally elongated in shape, although occasionally flattened in some species. The rostrum terminates behind in a point which sometimes bears a protuberance

*Ciroceratites emmerici, an ammonite of the Lower Cretaceous; Castellane, France. Diameter, 9.2 cm*

in the atractitides a proostracum is thought to have been present. In the later Jurassic belemnites the phragmocone is much smaller while the rostrum is more developed, and notable among these are the family of belemnitides, which furnish a number of guide species to Jurassic and Cretaceous rocks. In the belemnoteuthides, which were already present during the Upper Triassic but were principally to be found in the Jurassic and Cretaceous periods, the proostracum begins to develop and the phragmocone tends to bend over. Finally with the neobelemnitides are grouped forms that lived from the Cretaceous to the Tertiary period. These indicate a transition to the sepioids and tend to bear a phragmocone whose curve is even more pronounced. At the beginning of the Tertiary era, in the Eocene period, the suborder of belemnoids becomes completely extinct, giving place to the sepioids and teuthoids, which probably derived from it.

Members of the suborder of sepioids include the modern cuttlefish and some similar fossil

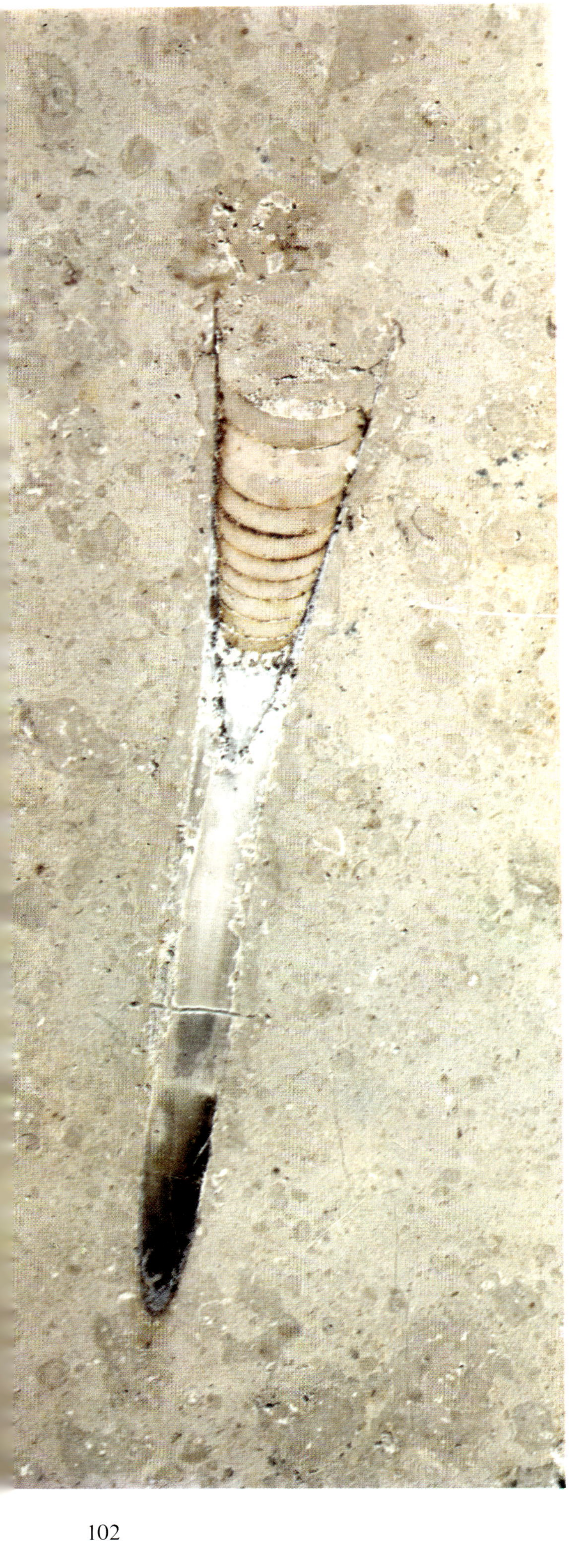

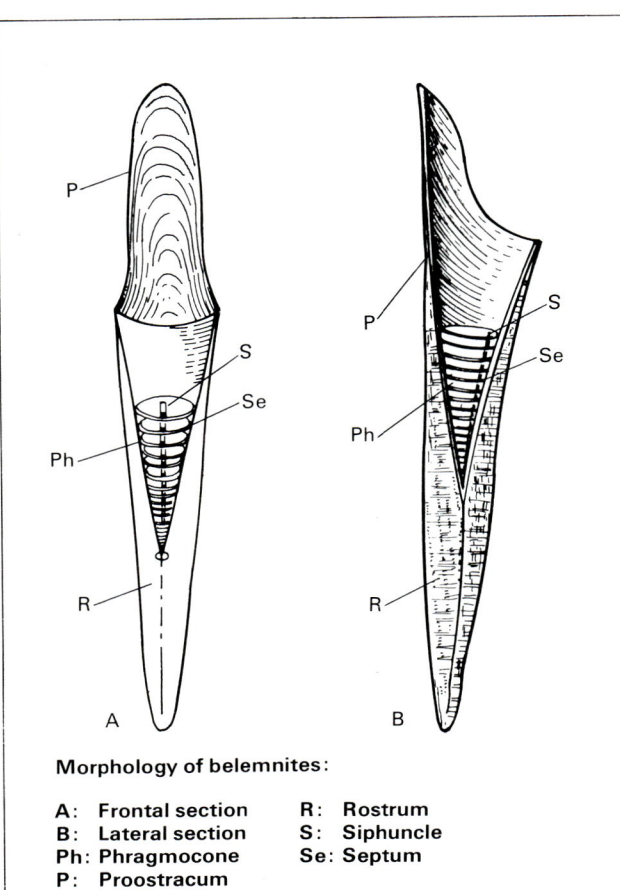

**Morphology of belemnites:**

A: Frontal section
B: Lateral section
Ph: Phragmocone
P: Proostracum
R: Rostrum
S: Siphuncle
Se: Septum

*Far left: Section of a belemnite in a rock of the Middle Jurassic; Feuchtwangen, Franconia, Germany. Clearly visible in the illustration are the elongated guard made up of crystals of calcite and, in the upper part joined to this, the chambered phragmocone. Length of the original, 7.5 cm*

forms. Among these are the Spirulirostridae, which lived from the Eocene to the Miocene period and in which the phragmocone shows a strong tendency to twist, and the Spirulidae, which have a spiral phragmocone. These last, which lacked a siphon, have been found in fossil form only in the Italian Miocene, and the sole descendants living today are confined to the oceanic waters at a depth of 3250 to 6500 feet. The Sepiidae, to which the common cuttlefish belongs, appeared in the Eocene period; their shell consists of an extremely small rostrum, a fully developed phragmocone that forms the central spongy part, and a large proostracum.

The suborder of teuthoides however includes the calamaries, whose shell has a proostracum that is highly developed in relation to the rostrum and phragmocone. These appeared during the Liassic period and have remained virtually unchanged down to our day.

Before concluding this chapter on the molluscs we should mention the octopods, an order of cephalopods to which belongs the genus *Octopus*. They first appeared in the Upper Cretaceous period with the genus *Palaeoctopus*, discovered in one single specimen for which a separate suborder was established. The suborder of polypodoids (Polypodoidea), has as a member the genus *Argonaut*, the female of which has a fragile flat spiralled shell.

*Right: Section of* Megateuthis giganteus, *a belemnite of the Middle Jurassic; Goslar, Germany. Length of the original, 12.5 cm. Note the well-developed guard and, in the upper part, the conical chambered phragmocone, which still preserves part of the conotheca. Far right: The greatly enlarged phragmocone of an actractitid,* Actractites indunensis, *of the Lower Jurassic; Furlo Pass, Italy. The separation between the various chambers is clearly visible in this specimen*

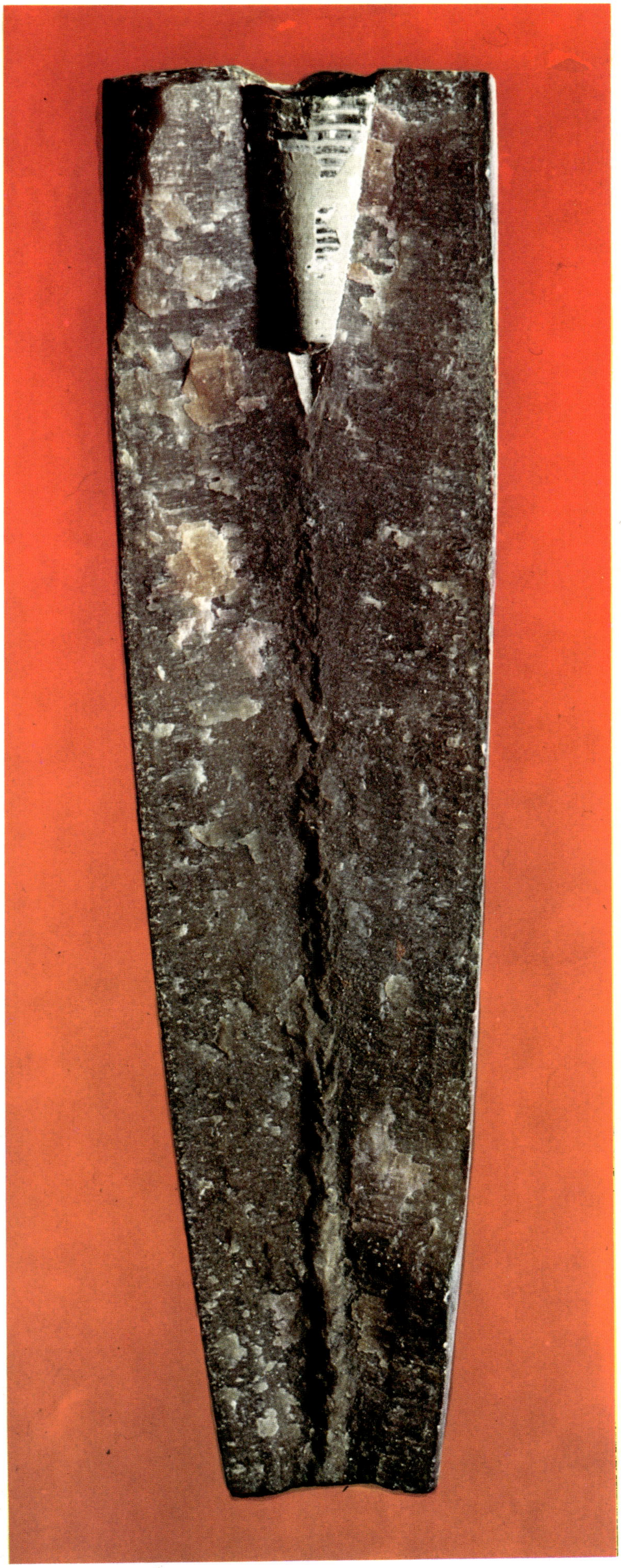

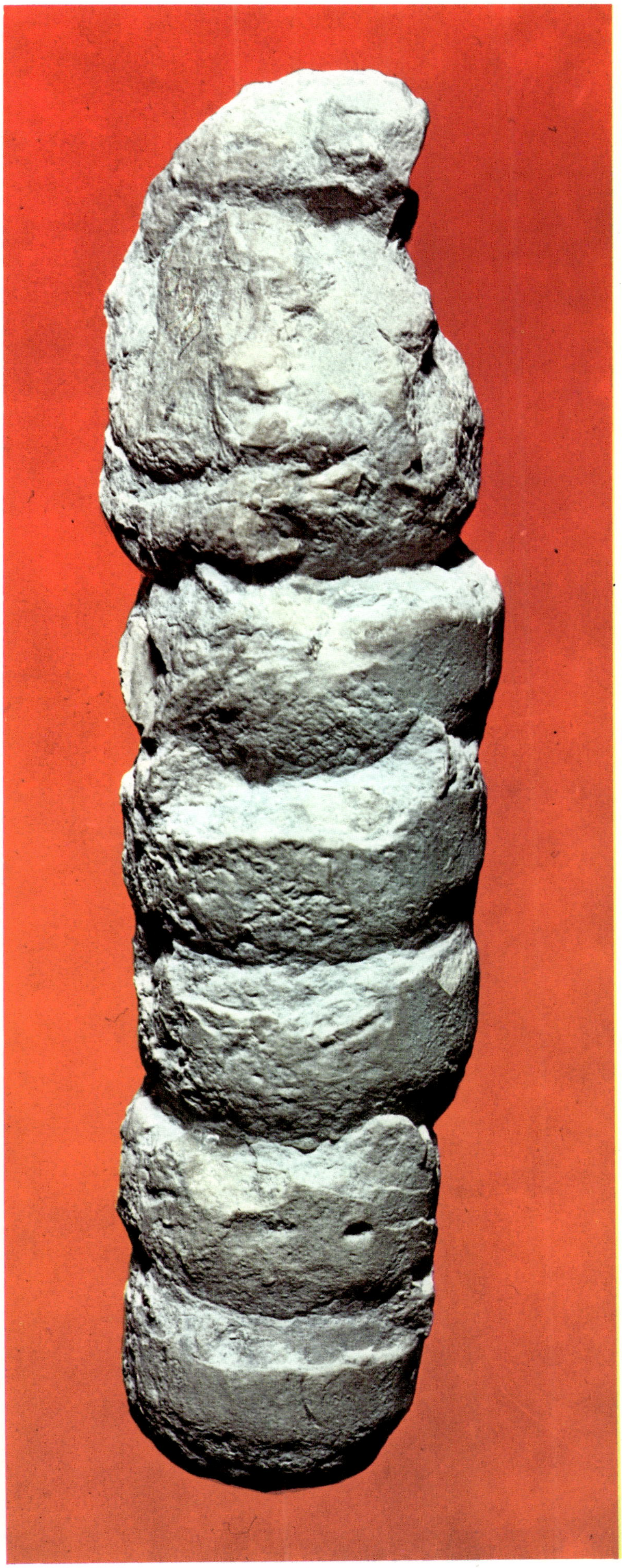

*Left: Interior of the shell of a primitive teutoid,* Plesioteuthis prisca; *Upper Jurassic, Eichstätt, Germany. Length of the original, 20 cm. The depression visible at the centre is the imprint of the ink-sac. Right: Typical guard of a belemnite,* Belemnites sulcatus. *Middle Jurassic; Erice, Sicily (×2)*

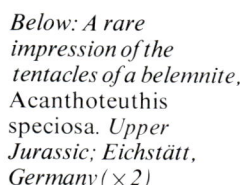

*Below: A rare impression of the tentacles of a belemnite,* Acanthoteuthis speciosa. *Upper Jurassic; Eichstätt, Germany (×2)*

# The Echinoderms

The echinoderms (Echinodermata) are an exclusively marine group of animals that have very well-defined characteristics. They are subdivided into a number of different classes, some of which had evolved even before the Lower Cambrian period and were extinct before the end of the Palaeozoic era, while others, with a slightly shorter evolutionary history, are not much younger, but have a large number of descendants still living. The most common of the echinoderms that still populate the seas today are the crinoids, the stelleroids, the echinoids and the holothurians, better known under the less technical names of sea lilies, starfish, sea urchins and sea cucumbers.

Typically, the echinoderms possess a dermal skeleton made up of calcium carbonate plates with characteristic pentaradial symmetry, and by an internal water supply system that permits a water circulation through the body through five radial canals. They are provided with an oral aperture and an anal aperture which are sometimes situated on opposite sides of the animal, and sometimes on the same side of the body. In the latter case, a striking bilateral symmetry is evident, particularly in the more mobile echinoderms.

In many echinoderms the body is divided into five radial extensions, which assume the appearance of arms in the stelleroids, of particular fixed areas in the echinoids, or of internal compartments in the holothuroids. Such extensions, which are known as ambulacral areas, follow the path of the water supply canals and are marked on the skeleton by five radial furrows, formed by porous plates, along which a flow of water laden with particles of food moves towards the mouth. The ambulacral areas are separated by groups of aporous plates that constitute the inter-ambulacral walls.

To the collector of fossils, the most interesting aspect of these animals is the dermal skeleton, which encloses the internal organs and is made up, as we have said, of calcareous plates. These are frequently ornamented with spines and tubercles, and are either articulated one with another or – as in the majority of cases – tightly

| Phylum | Class | Age |
| --- | --- | --- |
| Echinodermata | Cystoidea | Ordovician-Devonian |
| | Blastoidea | Silurian-Permian |
| | Edrioasteroidea | Cambrian-Carboniferous |
| | Crinoidea | Ordovician-Recent |
| | Stelleroidea | Ordovician-Recent |
| | Ophiocistoidea | Ordovician-Devonian |
| | Echinoidea | Ordovician-Recent |
| | Holothuroidea | Devonian-Recent |

*A typical crinoidal limestone: parts of the stems of* Encrinus *species in a stone of the Middle Triassic from Germany (slightly enlarged)*

joined together to form a firm protective covering. These plates are formed in the living animal by a network of spicules welded together, with their empty spaces filled with organic material. In the fossils the original aragonite of the spicules is transformed into calcite, which is much stronger, while the empty spaces are filled with the same mineral which, on crystallizing, gives rise to a robust skeleton that is usually well-preserved.

The echinoderms are particularly important for their use both as guide fossils and as accurate ecological indicators. For the echinoderms are strictly marine animals, only rarely adapted to life in estuarine waters, and it is thus logical that discovery even of one single plate should be a certain indication of marine surroundings. In addition, many species are adapted specifically to certain conditions on the sea bed, and are thus useful in estimation of the depth of the sea at the time that they were deposited.

In the same way, echinoderms have played a significant role in marine ecology. Many large deposits of limestone on the sea bed are the result of decomposition of crinoid skeletons, while large numbers of rock-boring urchins have had an important erosive influence on coastlines, and yet other echinoderms are important scavengers, or are themselves sources of food.

*Facing page: An association of some specimens of* Pentacrinus fasciculosus, *a crinoid (subclass Articulata) of the Lower Jurassic; Holzmaden, Germany (lifesize)*

*Below: A calyx of* Pentremintes godoni, *a blastoid of the Carboniferous; Huntsville, Alabama. Height of the original, 1.2 cm. The complex structure of the ambulacral area is evident in the illustration*

The phylum of the echinoderms, which is usually divided into Pelmatozoa, or fixed echinoderms, and Eleutherozoa, the free echinoderms, appeared suddenly in the Cambrian period, when many species have already attained specialized forms. These were members of a number of classes, among them the cystoids (Cystoidea) and edrioasteroids (Edrioasteroidea), which became extinct well before the end of the Palaeozoic era, and the holothuroids (Holothuroidea), which are today the only survivors of this age. For this reason the origin of the echinoderms, which is almost completely obscure, has been traced back to Precambrian times when, on the basis of embryological studies on living forms, it would appear that they derived from segmented worms through some unknown intermediate stage.

In the following period, the Ordovician, about 70 million years later, all the other groups appeared. Some of these, like the blastoids (Blastoidea) and ophiocystoids (Ophiocystoidea), became extinct in the Palaeozoic era, while others, like the crinoids (Crinoidea), stelleroids (Stelleroidea) and echinoids (Echinoidea), are to be found today after an evolution that has lasted for some 430 million years.

Of the various classes quoted in the table at the start of this section, we shall deal only with those that are of the greatest interest to a collector, either because they are very common or because they are more widespread, and thus more important to collectors, than others.

**Edrioasteroidea**

The edrioasteroids, echinoderms with a globular or flattened body, lived from the Cambrian to the Carboniferous period, attached to the sea bed by means of the base part, without the aid of a peduncle. Its flexible sheath was formed by irregular plates and it carried on the upper part five sinuous 'ambulacral areas' formed by very small and regular plates, at the centre of which was the mouth. The anal aperture, situated between the two ambulacral areas, was covered by a pyramid of small plates.

**Cystoidea**

The cystoids, echinoderms that are now extinct, lived from the Middle Ordovician to the Devonian period. They attached themselves to the bottom of the sea by means of a peduncle, formed by superimposed plates, that had inferior terminations similar to roots. At its upper extremity the peduncle supported a spherical sheath, which was slightly elongated or flattened, and could reach a maximum height of 40 centimetres (16 inches). This sheath was made up of an association

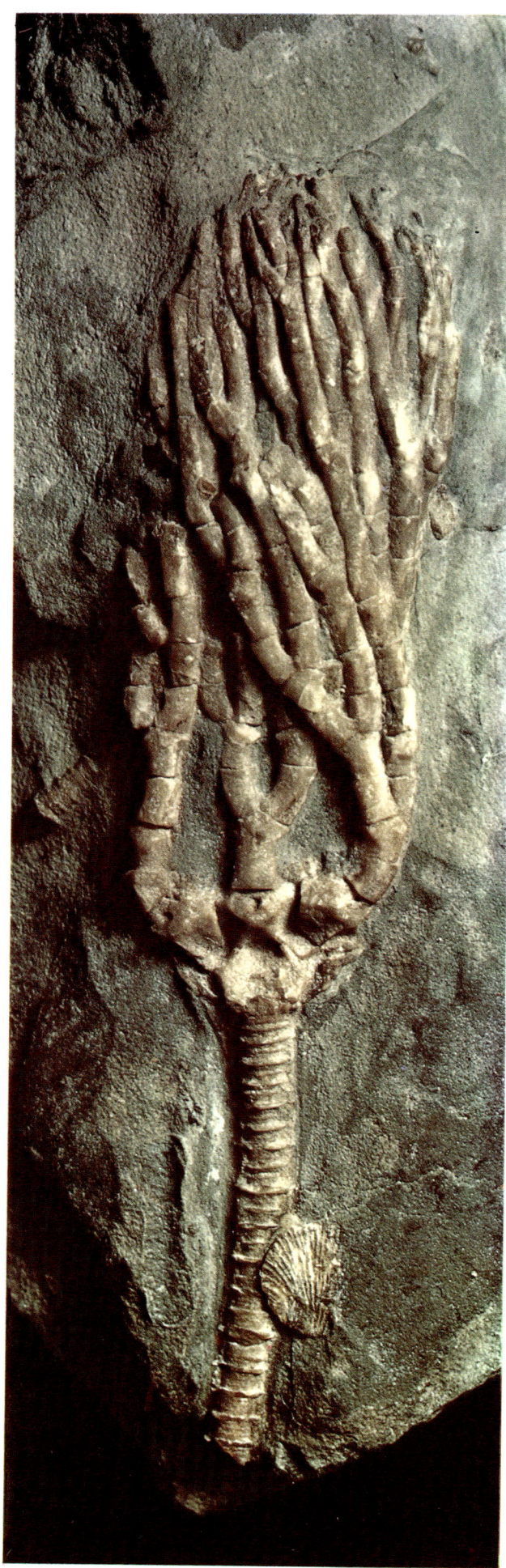

of hexagonal, pentagonal or irregularly shaped plates, arranged in alternate circles. The plates had either individual pores, a pair of pores, or rhomboid pores on each side of the suture line between two plates, and joined by horizontal channels. The classification of the cystoids is based upon the form and position of these pores.

The oral and anal apertures of the cystoids are to be found on the upper part of the sheath. In the most evolved forms, simple or bifurcated ambulacral areas formed by small plates covered by other smaller plates, radiate from the mouth. On the edges of each ambulacral channel are thin articulated appendages, known as 'horns', that cause the animal to look like a flower, in the same way as the more modern crinoids.

## Blastoidea

The blastoids are a class of echinoderms that appeared in the Silurian period and became extinct before the end of the Permian period.

The blastoids lived attached to the sea bed by means of a flexible peduncle, formed by a series of superimposed plates, which possessed root-like ramifications. At the upper end of the peduncle was an ovoid sheath or calyx, composed of three annular circles, each with five large plates, closed at the top by a flattened disc in the centre of which was the mouth. From this radiated five ambulacral zones similar to the petals of a flower, each formed by four series of small regular plates. These communicated with a very complex internal apparatus, which probably had both a respiratory and a reproductive function. Around the mouth were to be seen five holes, one of which was the anal aperture. The horns projected from the marginal plates of each area. In moving, these produced a flow of water, and thus procured food that the animal needed.

## Holothuroidea

The holothuroids, a well-known group that is very widespread in our seas today, constitutes the oldest class of living echinoderms, for its first members were already well distributed in the seas of the Cambrian period. They are differentiated from other echinoderms through the lack of a continuous skeleton: the body is in fact covered by a resistant envelope, inside which are distributed a large number of variable spicules, known as sclerites. These are the only parts of the animal that fossilize easily, and because of their microscopic size they are collected and studied by methods appropriate for the microfossils. Impressions of the complete animal are very rare, having only been found in terrains that allow the preservation of the most delicate structures.

*Far left:* Cyathocrinus goniodactylus, *a crinoid (subclass Inadunata) of the Silurian; Dudley, Britain. Height of the original, 5.3 cm*

*Right: A complete specimen of* Encrinus liliiformis, *a crinoid which is the best index fossil for the marine levels of the Middle Triassic; Württemberg, Germany. Length of the original, 15.7 cm*

Right: Articulated
peduncles of a crinoid
of the Middle Triassic,
Encrinus sp., Germany.
( ×2 approx)

Left: Two other
specimens of
Encrinus liliiforma,
from the Middle
Triassic in Germany
(lifesize). Below:
A crinoidal limestone:
fragments of the stems
of Encrinus sp. in
Middle Triassic stone.
Melide, Ticino,
Switzerland

## Crinoidea

The crinoids are one of the most interesting groups in the echinoderms, both because of the great variety of species that have derived from them over the geological eras, and because of the large numbers of species found.

The crinoids are strictly marine animals which generally live attached to the bottom by means of a long peduncle, although some species are known that became free-moving as adults, having been fixed during development. These lack a peduncle, and carry on the lower part of the calyx a number of tendrils, with which they can adhere to the solid substratum.

A complete crinoid consists of three distinct parts: the calyx, which encloses the vital organs; the arms, which are used to convey the particles of food to the mouth; and the peduncle or tendrils, which serve to fix the animal to the sea bed. The peduncle is formed by a number of columnar plates superimposed on each other which are fre-

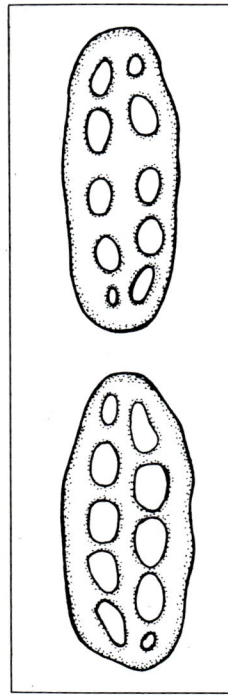

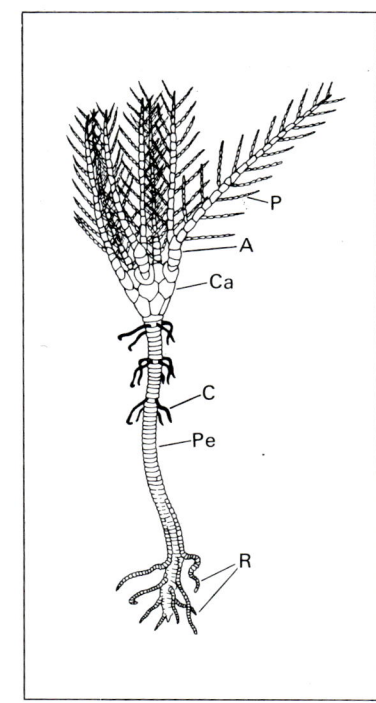

*Far left: Two specimens of the sclerites of Holothuroidea, much enlarged. Near left: Nomenclature of the crinoids: A. Arms; C. Cirri; Ca. Calyx; P. Pinnulae; Pe. Peduncle or stem; R. Roots or 'holdfast'*

*Left:* Saccocoma pectinata, *a crinoid without a stem, which is said by some authorities to be a larval form. Upper Jurassic; Solnhofen, Germany. Size of the original, 3.1 cm*

quently to be found on their own in sedimentary rock. Their shape varies remarkably from species to species: they may be circular, square, elliptical, pentagonal or stellate: some have a smooth surface, others are covered with radial lines, and yet others by petal-like designs. In the centre of each plate is a hole, which corresponds to a canal that runs through the whole length of the peduncle. At its lower end there are appendages in the form of roots, or – more rarely – in the form of an anchor, which facilitate their fixing to the terrain. At the upper end of the peduncle is the calyx, composed of circles of pentagonal plates, either fused together or articulated. The calyx is generally shaped like a cup, hollow internally, and containing the vital organs. These are protected from above by a circular cover that forms the oral surface, at the centre of which is the mouth. From this originate five ambulacral canals that join to the arms, which are formed by brachial plates disposed in a single or double series. These are frequently ramified, in order to increase the catchment area for food. Each brachial plate carries on its ventral side an appendage (pinnule) along which the ambulacral canals are continued.

The crinoids were already quite developed at the time of their first appearance during the Lower Ordovician period, and so it is thought that their origin must be sought further back in time. The crinoids are grouped into four subclasses. Of these, the subclasses Camerata, Inadunata and Flexibilia became extinct before the end of the Triassic period, having appeared during the Middle Ordovician. The subclass Articulata, which groups together virtually all the Post-Palaeozoic and recent genera, had its origin in this period.

In the subclass Camerata are to be found crinoids whose plates are closely welded together, forming a rigid calyx, and where the mouth and ambulacral canals are covered by a vaulted roof

*Euzonosoma sp., a stelleroid of the Lower Devonian. Bundenbach, Germany (×2.2)*

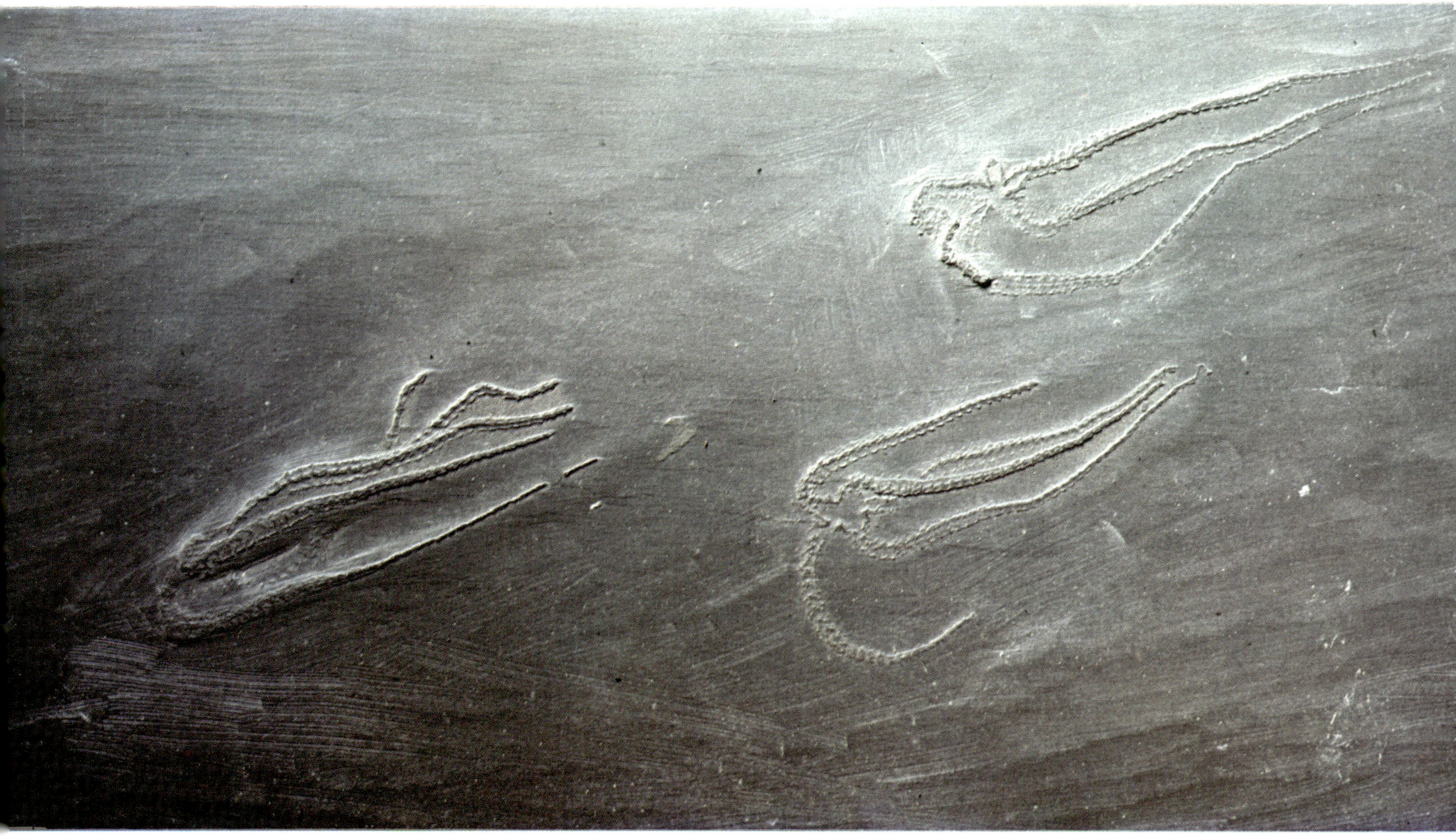

Above: Three specimens of Furcaster palaeozoicus, *a stelleroid of the Devonian. Bundenbach, Germany (slightly enlarged). The position of the arms indicates that the animal was deposited on the seabed under the influence of a strong current flowing from the left. Left:* Ophioderma egertoni, *an ophiuroid of the Lower Jurassic. Lyme Regis, Britain (size of the specimen, 3.1 cm)*

formed by further plates. The arms consist of one or two series of plates and bear well-developed pinnules. These animals were alive from the Middle Ordovician to the Permian period.

The subclass Inadunata groups together some genera to be found from the Lower Ordovician to the Triassic period; these include some of the oldest and most primitive crinoids. In these organisms, too, the calyx is formed of rigidly welded plates, and the mouth and ambulacral canals are covered by large plates. The single Mesozoic group attributed to this subclass is the family Encrinidae, whose genera have played an important part in the formation of some Triassic alpine sediments. The most common genus is *Encrinus,* and the species *E. liliiformis* is typical of the Middle Triassic, and is found as complete calices or isolated articulations. *Encrinus* has a depressed calyx, supported by a long peduncle made up of uniform circular articulations, ornamented by radial lines, with long arms formed by a double series of plates.

The subclass Flexibilia includes the crinoids whose calyx is formed by movable plates. These comprise a few genera that lived between the Middle Ordovician and the Permian period.

The subclass Articulata includes the majority of the Post-Palaeozoic and present-day crinoids, which have a calyx formed by articulating plates with an uncovered mouth, and ambulacral canals and arms made up of a single series of plates. Numerous genera and species are used as guide fossils, principally in Mesozoic terrains. The best known of these is the genus *Pentacrinus,* and the pentagonal articulations of its stem are to be found in vast numbers in the Triassic and Jurassic layers. Complete examples of this genus, which has a greatly reduced calyx and ramified arms, are found in the Upper Liassic layers of Holzmaden in Württemberg. These may be as much as several yards across.

Finally, the family of Comatulides which appeared in the Liassic era is still to be found today. Members of this family are attached by a peduncle in the juvenile stage, and free in the adult form.

### Stelleroidea

The stelleroids, which include the starfish and the ophiuroids that still exist today, are free echinoderms with a body formed by a central disc from which spring five arms. In some cases there may be more arms, and in exceptional cases, this has reached as many as 40. Inside the central disc are located all the organs of the animal. These continue along the arms to permit the animal to

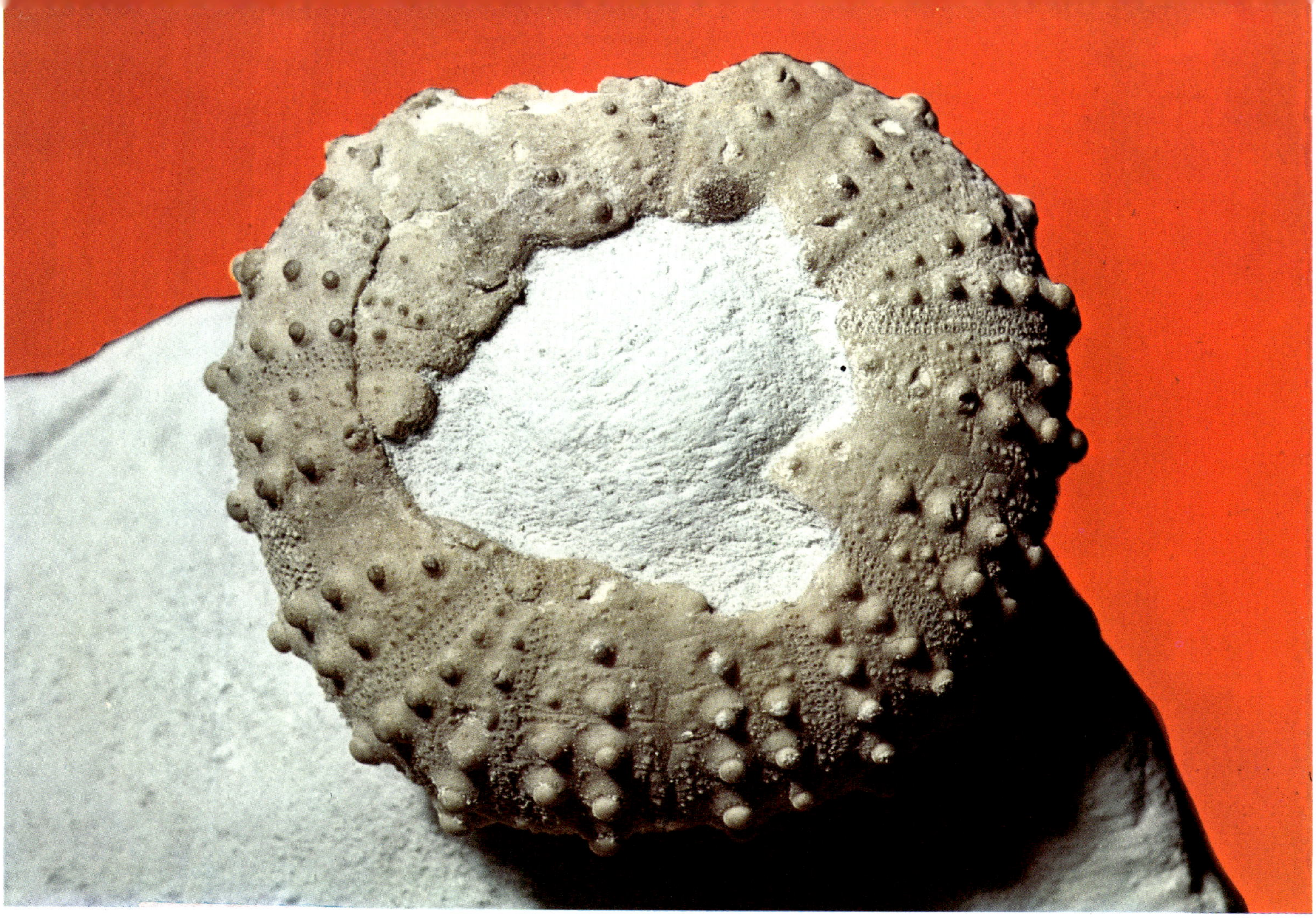

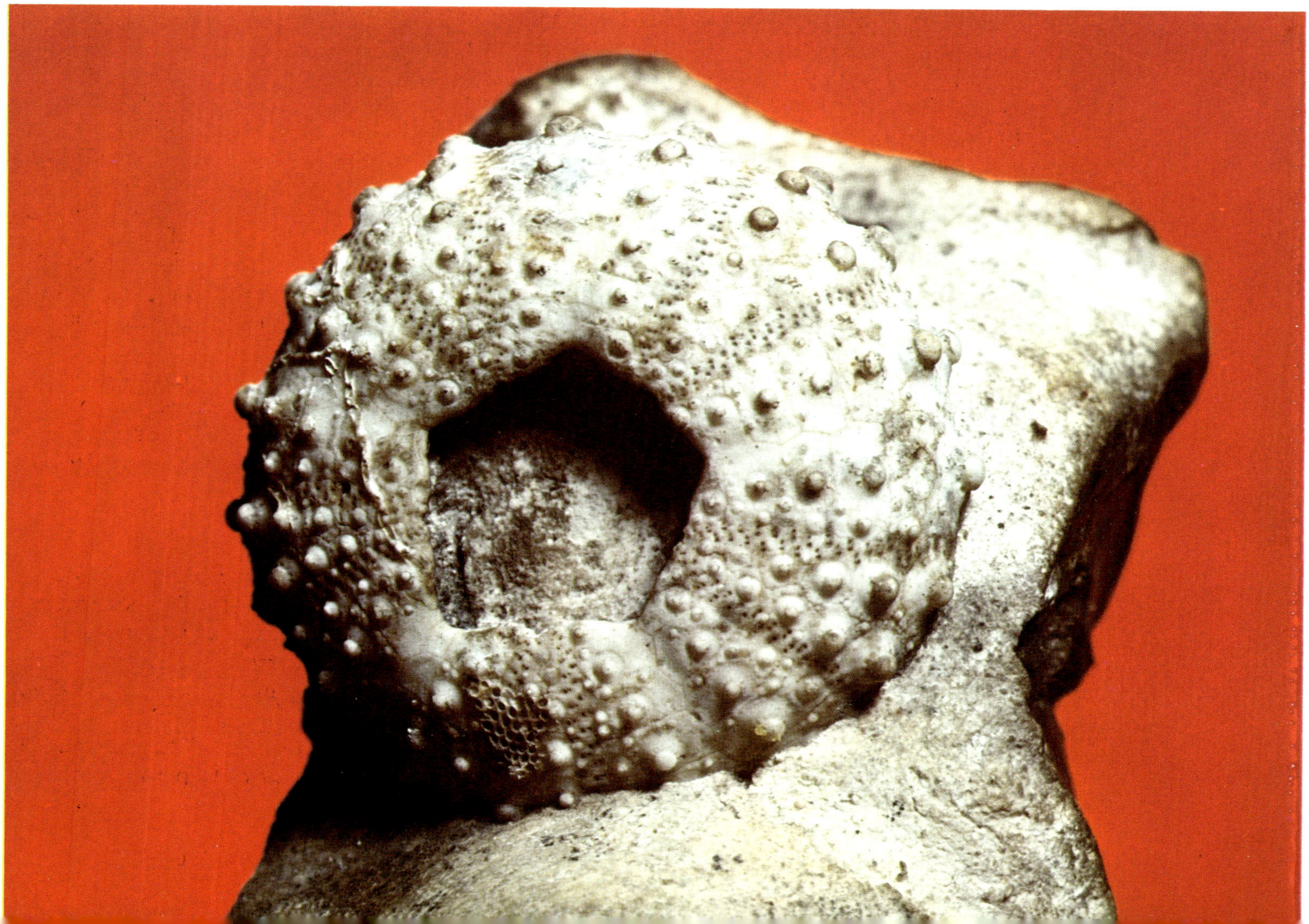

*Above:* Stenonia tuberculata, *an echinoid of the Irregularia. Upper Cretaceous; Monte Lessini, Italy. Size, 4.3 cm. Above left:* Cyphosoma koenigi, *one of the Regularia of the Upper Cretaceous, Island of Rugen. Apical view, diameter, 4.3 cm. Below left: Another specimen. Dover (×4 approx)*

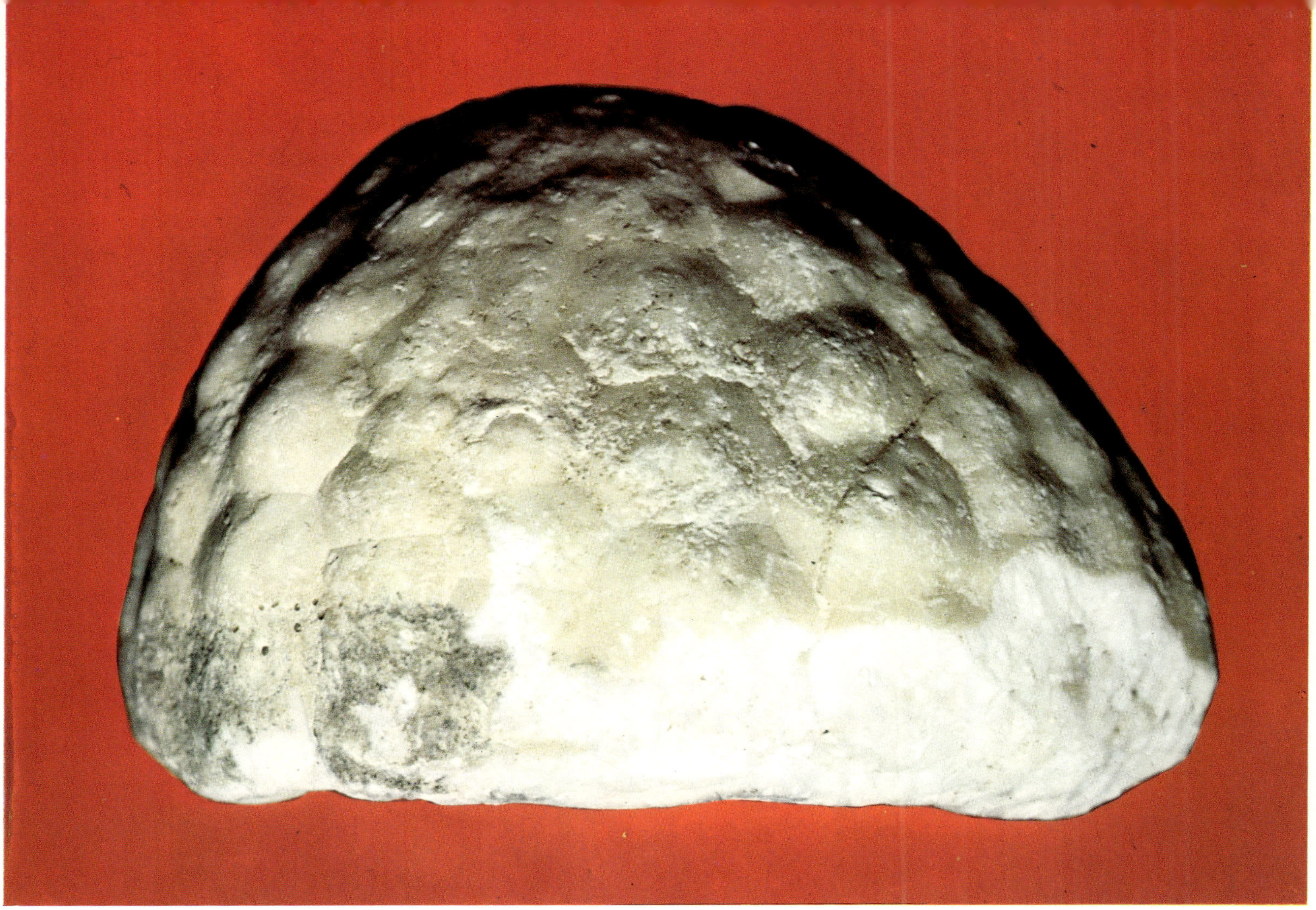

regenerate from a single fragment. In the lower part of the disc are the anal aperture and the mouth, from which come five ambulacral canals that run through the whole length of the arms. The skeleton, which may be divided into ventral and dorsal parts, is formed by numerous free plates held together by a tegument. It is because of this structure that it is rare to find these echinoderms in the fossil state, for the decomposition of the body after death frees the plates, which are commonly found on their own in the sediments.

The stelleroids live, as in the past, at all depths and in all latitudes; they are, however, of little palaeoecological significance. They appeared during the Lower Ordovician period, with the subclass somasteroids (Somasteroidea), which includes the most primitive representatives with a roughly pentagonal body. In the Upper Ordovician period the subclass ophiuroids (Ophiuroidea) appeared, together with the subclass of the asteroids (Asteroidea), to which the classic starfish, still alive today, is attributed.

*Right: Nomenclature of the Echinoidea. (A. Transverse section; B. Apical view.) A. Ambulacral area; An. Anus; Sp. Spines, M. Mouth; Ac. Ambulacral canal; Hc. Hydrophoric canal; S. Sex organs; Te. Test or exoskeleton; Ia; Interambulacral area; P. Pedicles; Mp. Madreporic plate; Vs. Visceral sac*

**Echinoidea**

The echinoids or echinids, better known as sea urchins, are echinoderms whose body is enclosed in a shell, lacking arms and peduncle, and composed of numerous regular plates. This shell may vary enormously in shape. It may be spherical, hemispherical, conical or discoid. In the living animal this shell encloses the soft organs, among them the twisted visceral sac which communicates with the exterior through the anal aperture above and through the mouth below. The position of these two apertures determines the symmetry of the echinids, which are accordingly divided into two groups: the *regular* echinids, which have pentaradial symmetry, with the two apertures situated at opposite poles of the shell, and *irregular* echnids, with bilateral symmetry due to the migration of the anal aperture along an interambulacral area.

Also inside the shell is the water supply system,

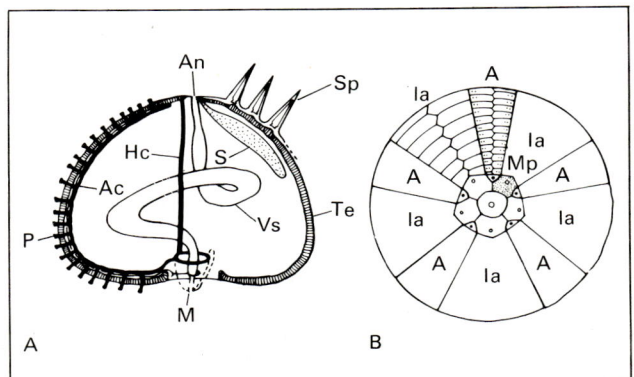

Micraster coranguinum
*an irregular echinoid
of bilateral symmetry,
with an ambulacral
area almost like the
petals of a flower.
Upper Cretaceous;
Paris Basin.
Size of the original,
5.5 cm*

Holectypus depressus,
*an irregular echinoid
of the Upper Jurassic;
Peterborough, Britain.
Diameter of the
original, 3 cm*

Clypeus ploti, *an irregular echinoid with a petaloid ambulacral area and bilateral symmetry. Middle Jurassic; Liesberg, Switzerland. Size of the original, 8.5 cm*

which consists of a vertical tube that leaves the apex near the anal aperture, to open out lower down into an annular tube, which runs round the mouth. From this radiate five ambulacral canals which pass through the walls of the shell from the plates of the ambulacral areas, and the animal puts extensions that are used for loco-motion and for respiration through the holes in these plates.

As in all animal groups, the only part that fossilizes well is the calcareous shell, and it is on the structure and shape of this that the palaeon-tological classification of the echinids is based. This skeleton is formed, as we have said, by a large number of calcareous plates closely joined to each other. They lie in different vertical series and form ten zones, or areas, each made up of two series of plates. Five ambulacral zones formed by perforated plates alternate with five inter-am-bulacral zones made up of plates without pores.

The mouth is situated at the lower end and is composed of different calcareous pieces that constitute the complex so-called 'lantern of Aristotle', or masticatory apparatus. Some important systematic subdivisions are based upon the presence of this apparatus, which is sometimes lacking, and upon the appearance of its edges. As we have said, it is possible for the anal aperture to be displaced on the upper side towards the lateral edge, or even the lower face, of the shell.

Another important feature is that the interambulacral plates of the regular echinids have a long clearly marked spine of variable shape and size. This is inserted into a tubercle surrounded by an annular depressed zone and by a ring of smaller tubercles. In the depressed zone the muscles that move the spine are inserted, and are themselves protected by smaller spines and articulated plates on the tubercles of the ring. Examples of fossil echinids complete with spines are rare; much more frequently the shells are intact and the spines found separately.

The first echinids appeared in the sea during the Ordovician period, derived, it is believed, from the cystoids. These were the palaeoechinids,

a group of primitive echinids that lived for the whole of the Palaeozoic era, and finally became extinct in the Permian period, when the regular 'modern' echinids appeared. They only began to be widely distributed in the Triassic period, however, with the cydarids, a group very similar to the sea urchins of today, whose species are used as guide fossils for some levels of the Mesozoic period. The irregular echinids appeared rather later, during the Liassic period, and during the Cretaceous period attained such vast numbers that they provided a large number of guide species for this period. By the beginning of the Cenozoic era the regular echinids were diminishing in importance, while the irregular echinids took on a steadily growing importance.

In the same way, these animals also furnish the palaeontologist with precise clues as to the appearance of the sea bed in past eras. The regular echinids principally became adapted to life on rocky bottoms at different depths, while the irregular types generally lived on a sandy or muddy sea bed in shallow waters. Their discovery may thus be helpful to give a rough idea of the sort of sea bed that existed where they were found.

# The Graptolites

The graptolites are colonial animals that lived in the sea from the Middle Cambrian to the Lower Carboniferous period and disappeared completely during this period, leaving small carbonized bodies as traces of their existence.

Well known to all collectors, they have long represented a problem for palaeontologists, who have not been able to apply a satisfactory classification to them. This job has not been made easier by the fact that there are no living animals of a similar structure that would provide useful comparisons.

Thought at first to be of a vegetable or inorganic nature, the graptolites were next tentatively attributed to the cephalopods or the coelenterata, or were considered an isolated group within the animal kingdom. Only the discovery of specimens in a particularly good state of preservation has made possible a more detailed analysis of their structure, on the basis of which it has been established that the organisms belong to the stomochordates.

The stomochordates occupy a position within the animal kingdom between the invertebrates

| Phylum | Subphylum | Class | Order | Age |
|--------|-----------|-------|-------|-----|
| Chordata | Stomochordata | Enteropneusta | | Recent |
| | | Pterobranchia | Rhabdopleurida<br>Cephalodiscida | Cretaceous-Recent<br>Ordovician-Recent |
| | | Graptolithina | Dendroidea<br>Tuboidea<br>Camaroidea<br>Stolonoidea<br>Graptoloidea | Cambrian-Carboniferous<br>Ordovician-Silurian<br>Ordovician<br>Ordovician<br>Cambrian-Silurian |

*Above:* Dictyonema flabelliforme, *a dendroid graptolite. Ordovician; Germany (×2.5) Right:* Monograptus colonus, *a graptolite of the Silurian; Gutturu, Sardinia (×12)*

121

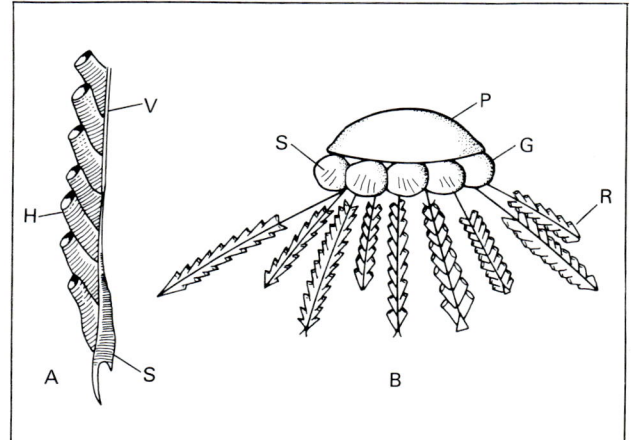

and the chordates (part of the vertebrates) and have only recently been found as fossils. In fact the stomochordates possess characteristics common to both the invertebrates and the chordates; like these last they have a series of branchial fissures in the pharyngeal region, and a nervous system in the dorsal position, but they lack the dorsal cord, this being replaced by a 'stomochord'.

The stomochordates comprise three groups of organisms which are taken to be classes: the Enteropneusta, the Pterobranchiates and the Graptolites.

Members of the first two classes are organisms such as *Balanoglossus* and *Rhabdopleura,* which are quite commonly found in the sea today. It is with the latter genus, which has come down to us from the Cretaceous period, that the graptolites show the greatest affinity: the rhabdopleura are animals that live attached to the sea bed in colonies equipped with a chitinous skeleton. The colonies are formed by primary horizontal tubes and secondary vertical tubes, open at the upper end, in each of which one individual lives. These tubes are made up of alternating semicircular chitinous rods in the primary tubes and in the older part of the secondary tubes, and by complete rings in the terminal part of the latter.

A very similar skeletal structure can be seen in the graptolites. These are also colonial animals, forming small colonies enclosed in a chitinous exoskeleton. The colony (or rhabdosome) is formed by superimposed small semi-annular segments. It originates from an empty conical chamber (sicula), which grows by budding. The end of the sicula extends to form the axis of the graptolite (virgula), a rigid stem, branched or simple, straight or curved, which carries on the side a series of small galleries that communicate with each other by means of a channel, each one inhabited, as in the rhabdopleura, by one individual of the colony.

*Association of* Monograptus lobiferus *and* Demirastrites triangulatus *(above left), both graptolites of the Silurian; Thuringia, Germany (×3 approx)*

*Nomenclature of the graptolites (A. Detail of a rhabdosome; B. A floating colony): G. Gonotheca; H. Hydrotheca; P. Pneumatophore; R. Rhabdosomes; S. Sicula; V. Virgula*

122

This general scheme is roughly the same in all graptolite colonies. The vast numbers of different species found in the rock layers of the Palaeozoic era has made the group extremely complex, though very useful to the palaeontologist, who, precisely on the basis of the variety, uses the graptolites as guide fossils to determine the age of the various sedimentary strata.

The class Graptolithina is divided into five orders, each with quite distinct characteristics.

Some are sessile, benthonic or epiplanktonic organisms. Among these are the dendroid graptolites, which have irregularly branched forms and a pedicle with which the rhabdosome adhered to the bottom of the sea. They are the oldest graptolites but also the longest-surviving, since they appeared before all the others in the Middle Cambrian and lived until the Lower Carboniferous period.

In the planktonic graptolites, however, the rhabdosome is simpler, sometimes formed by a single straight or curved shaft, and sometimes slightly branched. Many rhabdosomes of these animals may be associated, being attached to each other by means of an extension of the virgula on a disc (pneumatophore) in the form of a dome that must also have served as a flotation organ. At the base of some pneumatophores have been found some well-preserved small spheres containing a certain number of sicules. These were the small embryonic chambers from which the entire rhabdosome originated, and these spheres are thus known to be the reproductive organs of the animal.

The planktonic graptolites lived principally during the Ordovician and the Silurian periods, which as a result are often known as 'the graptolite era'.

*Below left: Monograptus chimaera. Silurian; Germany (much enlarged). Bottom right: Some specimens of Monograptus sp., in a thin section. Silurian; Flumini-maggiore, Sardinia (×11). The structure of the hydrotheca is clearly visible in the largest specimen*

# TRACES OF LIFE

In writing a book about the most common invertebrate fossils, one cannot completely ignore the traces of life left by the animals of the past, whether vertebrate or invertebrate.

By 'traces of life' we mean the eggs, the footprints, the tracks, the excrement, the embryos, the remains of food and of habitations. The study of such traces is a fascinating subject. They are con-

sidered to be true fossils and are the source of much useful information about the habits and the way of life of many animals now extinct.

The traces most commonly found in the sedimentary rocks are those that are now thought to be movement tracks made by invertebrates, although they were at one time thought to be vegetable imprints.

*Right:* Arthrophycus *sp., fossil traces believed to be tracks left by arthropods or worms. Upper Silurian; Kufra, Libya (lifesize)*

*Far left: The imprint of a tetrapod in the continental red sandstone of the Lower Permian; Nierstein, Germany (×1.5). Near left: A hyena coprolite discovered in the Quaternary deposits of Sicily. The analysis of coprolites (fossilized excrement) helps to determine the food of extinct animals and in many cases to identify the internal parasites which infested them. Right: Chondrites (the smaller traces) and Fucoides (the larger traces) on a Cretaceous rock of the Lombardy foothills (×1.5)*

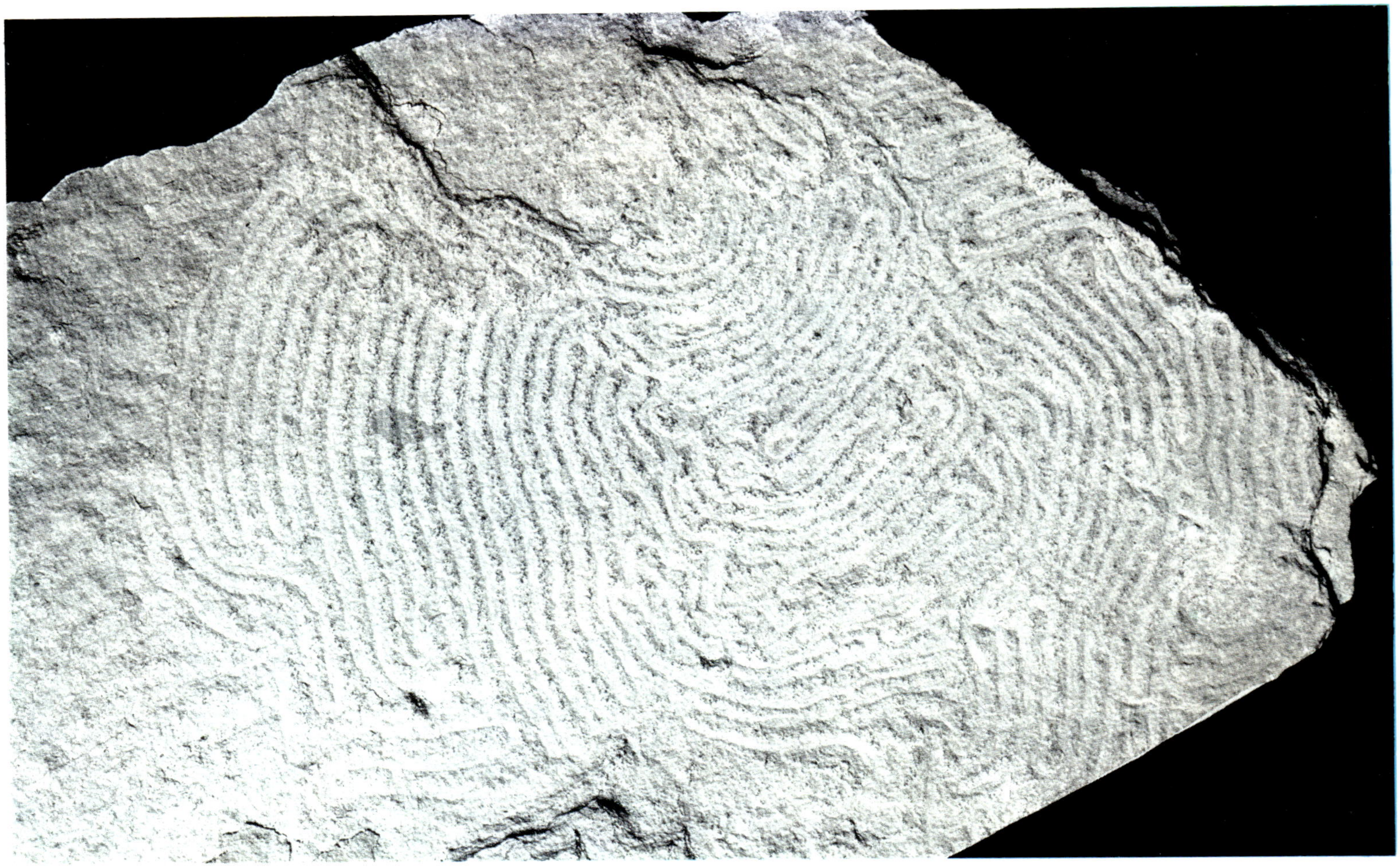

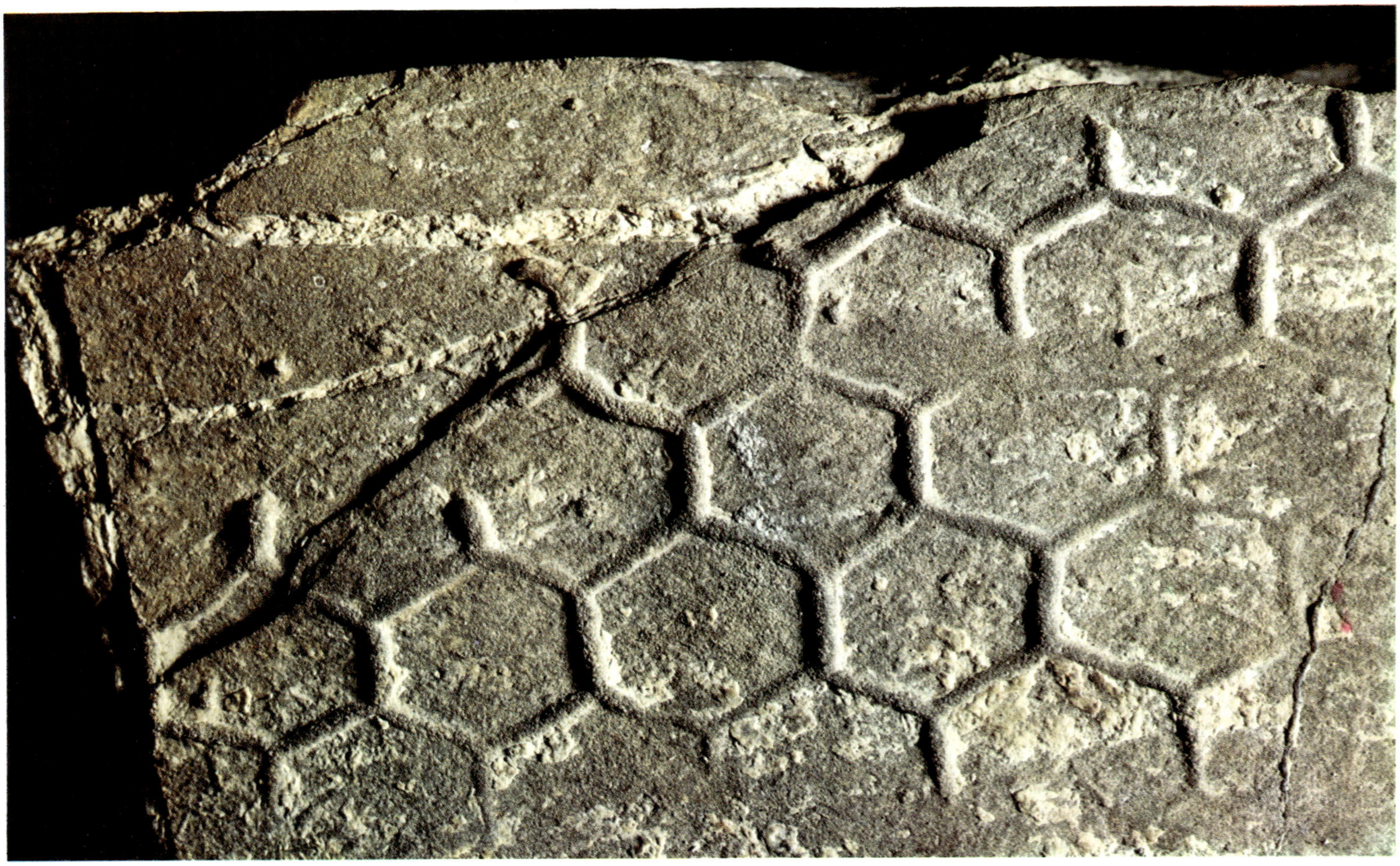

The *Chondrites* are impressions that were probably left by marine limivorous worms. They are in the form of irregular ramifications that represent the filling-in of small channels excavated by the animal in the mud. The *Fucoides* are filling-in structures of branched small tubes that look like plants, similar to the chondrites, and also due to the activities of worms. The Helminthoida are made up of a number of equidistant furrows, about two millimetres wide, often arranged concentrically. According to some authors they are due to limivorous worms, according to others they are tracks of shell-less gastropods which were not preserved. The *Paleo-dictyon,* found in terrains from the Ordovician to the Tertiary period, are strikingly different from all the foregoing, taking the form of a regular network of pentagonal, octagonal and hexagonal meshes. Nothing is known about these and they are considered by many to be of an inorganic origin. The *Zoophycos,* which present a bizarre leaf form, were also thought to be of an inorganic origin, but it has recently been discovered that these are due to sedentary annelids.

# Looking for fossils

In the chapter devoted to fossil protozoa we dealt with micropalaeontology, and the methods used in the collection and study of microscopic fossils. Here, on the other hand, we shall discuss the ways in which larger fossils should be collected and prepared.

The invertebrate fossils are generally found in all the compact or friable sedimentary rocks of marine, freshwater and continental origin. The fossils embedded in harder rocks, such as compact limestones or dolomites, are extracted only with difficulty. It is therefore important to inspect carefully the exposed surface of the rocks, since when they have been exposed to the air for a considerable length of time, fossils tend to come away very much more easily. A hammer and chisels of various dimensions should be used for extraction from a compacted rock. To find isolated fossils however one has to search carefully in the layers of detritus at the base of rock walls or along the courses of dried-up streams. The value of such specimens found is very variable, and because they have become dislodged and may have no trace of the original rock bed adhering to them, it is often not possible to identify the strata from which they came.

For the fossils in friable rocks, such as the clays, extraction is less difficult; a small hole made with a suitable hammer is usually enough.

Preparing the fossil means freeing it completely from the enclosing rock. This is generally done by mechanical methods, or – more rarely – by chemical means.

The mechanical methods consist in removing carefully the rock fragments round the fossil, with the aid of small chisels, a steel pick, dental drills, and, occasionally, the use of a few drops of very dilute hydrochloric acid when a fossil has to be extracted from limestone.

The chemical methods, on the other hand, aim to loosen the fossil from the rock with chemical reagents, when this is of a different type from the fossil that it contains. For calcareous rocks enclosing siliceous fossils, a 25 per cent solution of hydrochloric acid is used; this loosens the limestone, but leaves the silicate of the fossil intact. For calcareous fossils in non-calcareous rocks, hydrofluoric acid should be used – with the strictest possible precautions. Finally, calcareous fossils in calcareous rocks can be separated by heating the rock and then immersing it into cold water. Whatever method is used, however, it is important to remember that some fossils, particularly the more recent ones, are very fragile. Before they are examined, they should first be consolidated by means of a mixture of glycerin and gum arabic, or a solution of gum lacquer or of nitrocellulose. Extremely delicate specimens should ideally be kept in glass tubes lined with cotton wool.

# Bibliography

The classification of invertebrate fossils is frequently very difficult. While it is often simple enough to attribute a given animal to a class, an order or even to a family, it is a much more complex matter to know the genus to which the organism belongs. There are however some important books that cover the subject, and these will prove valuable, particularly to the amateur, who is attempting classification by genus.

The most important works, not only for problems of classification, but also for those who are interested in reading further about the subject, include:

Bather, F. A., *The Genera and Species of Blastoidea,* British Museum, London 1899.

Bisat, W. S., *The Carboniferous Goniatites of the North of England and Their Zones,* Proc. Yorks. Geol. Soc. **20**: 40-124, 1934.

Dalby, D. H., *Instructions to Young Geologists,* Museum Press Ltd., London 1957.

Elles, G. L., *The Graptolite Faunas of the British Isles.* Proc. Geol. Ass. London **33**: 168-200, 1922.

Evans, I. O., *The Observer's Book of British Geology,* F. Warne & Co. Ltd., London & New York 1949.

Howarth, M. K., *The Yorkshire Type Ammonites and Nautiloids of Young & Bird, Phillips and Martin Simpson,* Palaeontology, **4**: 520-37, 1962.

Kummel B., & Raup, D., *Handbook of Palaeontological Techniques,* Freeman and Company, San Francisco, London 1965.

Moore, R. C., *Treatise on Invertebrate Palaeontology,* Geol. Soc. of America and University of Kansas Press. Several volumes still to be published.

Moret, L., *Manuel de Paléontologie animale,* Masson, Paris 1958.

Piveteau, J., *Traité de Paléontologie,* Vols. I, II, III. Masson, Paris 1958.

Rhodes, F. H. T., Zim, H. S., & Shaffer, P. R., *Fossils,* Paul Hamlyn, London 1965.

Rudwick, M. J. S., *Living and Fossil Brachiopods,* Hutchinson University Library (Biological Sciences) 1970.

Shrock, R. R., and Twenhofel, W. H., *Principles of Invertebrate Palaeontology,* McGraw-Hill, New York 1953.

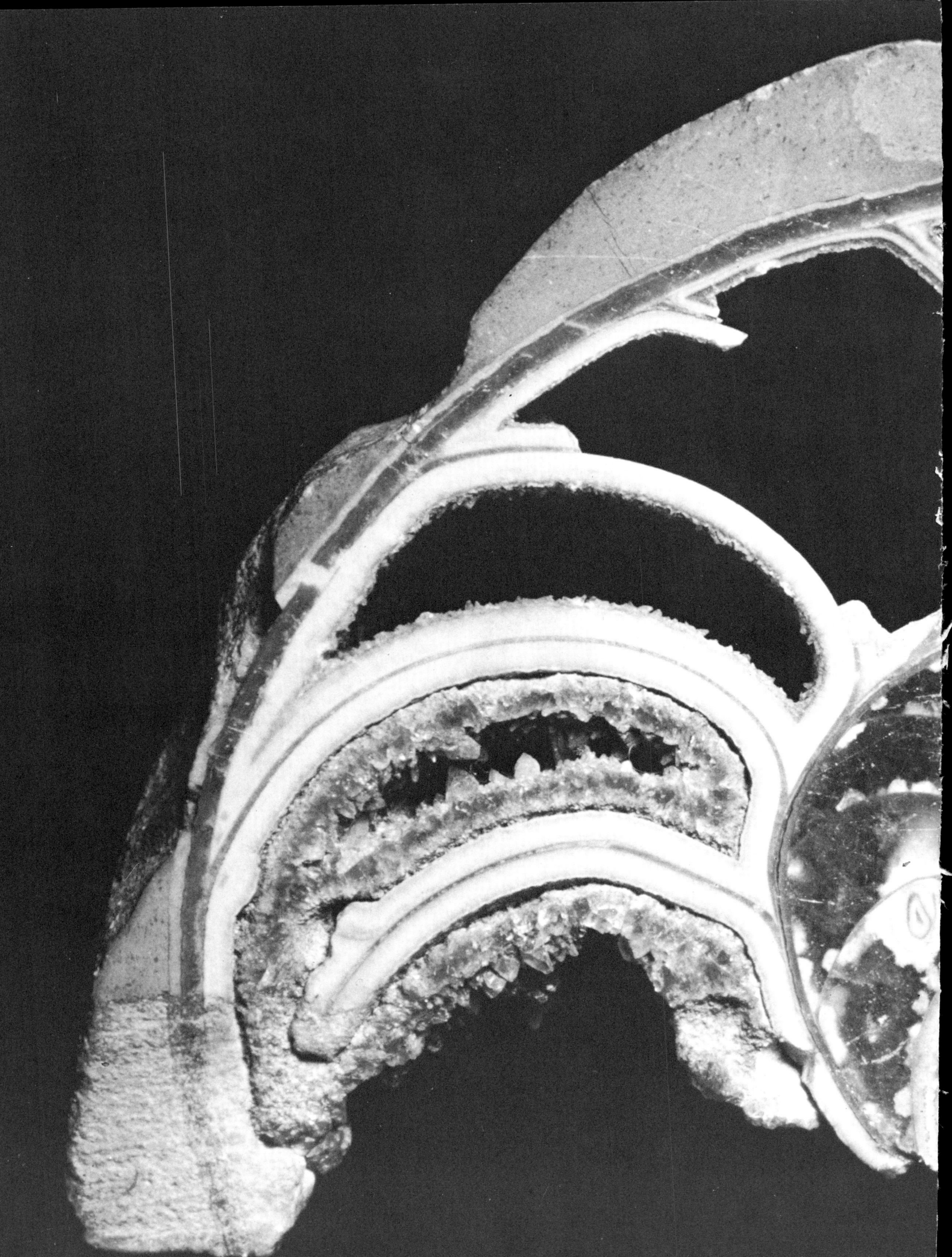